T0181133

Intelligent Systems Reference Library

Volume 88

Series editors

Janusz Kacprzyk, Polish Academy of Sciences, Warsaw, Poland
e-mail: kacprzyk@ibspan.waw.pl

Lakhmi C. Jain, University of Canberra, Canberra, Australia, and
University of South Australia, Adelaide, Australia
e-mail: Lakhmi.Jain@unisa.edu.au

About this Series

The aim of this series is to publish a Reference Library, including novel advances and developments in all aspects of Intelligent Systems in an easily accessible and well structured form. The series includes reference works, handbooks, compendia, textbooks, well-structured monographs, dictionaries, and encyclopedias. It contains well integrated knowledge and current information in the field of Intelligent Systems. The series covers the theory, applications, and design methods of Intelligent Systems. Virtually all disciplines such as engineering, computer science, avionics, business, e-commerce, environment, healthcare, physics and life science are included.

More information about this series at http://www.springer.com/series/8578

Jair Minoro Abe · Seiki Akama
Kazumi Nakamatsu

Introduction to Annotated Logics

Foundations for Paracomplete and Paraconsistent Reasoning

Jair Minoro Abe
Paulista University
Sao Paulo
Brazil

Kazumi Nakamatsu
University of Hyogo
Himeji
Japan

Seiki Akama
Kawasaki
Japan

ISSN 1868-4394 ISSN 1868-4408 (electronic)
Intelligent Systems Reference Library
ISBN 978-3-319-38686-7 ISBN 978-3-319-17912-4 (eBook)
DOI 10.1007/978-3-319-17912-4

Springer Cham Heidelberg New York Dordrecht London

Softcover reprint of the hardcover 1st edition 2015

Printed on acid-free paper

Springer International Publishing AG Switzerland is part of Springer Science+Business Media (www.springer.com)

Foreword

The present book constitutes an introduction to annotated logic, which is a kind of paraconsistent logic. This category of logic has a great theoretical relevance, especially in various domains of philosophy, the foundations of science, and mathematics. Annotated logic, in particular, also possesses a remarkable value for technology; for example, it is important in computer science, applied economics, database theory, and artificial intelligence.

The usual systems of paraconsistent logic can be viewed from two different perspectives: (1) As rivals of classical logic, for example when employed in certain formalizations of dialectics or in the foundations of quantum mechanics; (2) As logics complementary to classical logic, when, for instance, paraconsistent negation is seen as a weak type of negation, which is what happens in some applications to database theory.

In this volume, the second stance is practically assumed. The authors' central idea is to show how annotated logic can be applied as a tool to solve problems of technology and of applied science.

The text gives to the reader a clear view of the meaning of annotated logic and develops various significant applications of annotated paraconsistency. This is an excellent, well written, textbook, which discusses the principal traits of applied annotated logic. It will be of interest to pure and applied logicians, philosophers, and any person involved in the area of technology and applied science. The layman will also take profit from its reading.

Florianópolis
January 2015

Newton C.A. da Costa

Preface

Reasoning about incomplete and inconsistent information (also imprecise, paracomplete information) is an important subject in the fields of mathematics, philosophy, computer science, and Artificial Intelligence. To formalize such reasoning, *logic* plays an important role. This is because logic can precisely represent information and can derive useful information from given information as inferences.

Here, by logic we normally mean *classical logic*. It is known that classical logic is well established in the sense that its foundations, namely proof and model theory, have been fully studied. In addition, a lot of work on *automated theorem-proving* for classical logic has been done. The development of theorem-proving techniques has led to *logic programming* like Prolog.

However, classical logic is not suited to formalize incomplete and inconsistent information in our world. For instance, human reasoning is done based on incomplete and inconsistent information and involves some intensional concepts like knowledge, belief, and time. To overcome shortcomings of classical logic, various *nonclassical logics* have been proposed. They include modal logic, epistemic logic, tense logic, and others.

Annotated logics are systems of nonclassical logics for reasoning about incomplete and inconsistent information, originally proposed by Subrahmanian [149] as a foundation for *paraconsistent logic programming*. Later, da Costa et al. [62, 66] worked out formal aspects of annotated logics; also see Abe [1]. Since annotated logics were designed as a framework of reasoning about incomplete and inconsistent information, they can also be expanded for describing various types of common-sense reasoning.

We can view annotated logics as interesting systems of *paraconsistent logics*, which are logical systems for inconsistent, but nontrivial theories. Thus, annotated logics should be formally studied. In addition, there is a rich variety of applications of annotated logics.

This book is written as an introduction to annotated logics. The main objective is to provide logical foundations for annotated logics. We also discuss some interesting applications of these logics. The book includes the authors' contributions to annotated logics.

The structure of the book is as follows:

Chapter 1 gives motivations and the history of annotated logics as an introduction to this book. First, we discuss the importance of annotated logics in connection with paraconsistent logics. Second, we present the history of annotated logics by reviewing the literature on the subject.

Chapter 2 introduces the propositional annotated logics $P\tau$. We present a Hilbert style axiomatization of $P\tau$ and their semantics. We show some formal results including completeness.

Chapter 3 studies the predicate annotated logics $Q\tau$, which can be seen as a predicate extension of $P\tau$. Their axiomatization and semantics are considered. We also prove completeness and other metatheorems.

Chapter 4 discusses formal issues of annotated logics. We describe an algebraic semantics for $P\tau$ based on Curry algebras. We also discuss annotated set theory, annotated model theory, proof methods, and annotated modal logics.

Chapter 5 reviews some variants of annotated logics and related systems in the literature. Variants include fuzzy annotated logics, possibilistic annotated logics, inductive annotated logics, and structural annotated logics. We also compare annotated logics with related systems such as Labelled Deductive Systems and General Logics. Finally, we review systems of paraconsistent logics.

Chapter 6 discusses applications of annotated logics for various areas. After reviewing paraconsistent logic programming and generalized annotated logic programming, we survey promising applications to knowledge representation, neural computing, automation, and robotics.

Chapter 7 gives some conclusions with the summary of the book. It is possible to conclude that annotated logics are very interesting theoretically as well as practically. However, there are some future problems to be worked out.

We are grateful to Prof. Newton da Costa for helpful comments. We also thank Prof. John Fulcher for his suggestions.

January 2015

Jair Minoro Abe
Seiki Akama
Kazumi Nakamatsu

Contents

1 **Introduction** 1
1.1 Motivations 1
1.2 History 3

2 **Propositional Annotated Logics $P\tau$** 5
2.1 Language 5
2.2 Semantics 6
2.3 Axiomatization 13
2.4 Formal Results 16

3 **Predicate Annotated Logics $Q\tau$** 25
3.1 Language 25
3.2 Semantics 26
3.3 Axiomatization 28
3.4 Formal Results 29

4 **Formal Issues** 31
4.1 Algebraic Semantics 31
4.2 Annotated Set Theory 37
4.3 Annotated Model Theory 40
4.4 Proof Methods 48
4.5 Annotated Modal Logics 53

5 **Variants and Related Systems** 61
5.1 Fuzzy Annotated Logics 61
5.2 Possibilistic Annotated Logics 64
5.3 Inductive Annotated Logics 71
5.4 Structural Annotated Logics 74
5.5 Related Systems 95
5.6 Systems of Paraconsistent Logics 98

6 Applications . . . 111
6.1 Paraconsistent Logic Programming . . . 111
6.2 Generalized Annotated Logic Programming . . . 124
6.3 Knowledge Representation . . . 138
6.4 Neural Computing . . . 154
6.5 Automation and Robotics . . . 162

7 Conclusions . . . 175
7.1 Summary . . . 175
7.2 Future Problems . . . 176

References . . . 179

Index . . . 187

Chapter 1
Introduction

Abstract This chapter gives motivations and the history of annotated logics as an introduction to this book. First, we discuss the importance of annotated logics in connection with paraconsistent logics. Second, we present the history of annotated logics by reviewing the literature on the subject.

1.1 Motivations

Classical logic now holds a dominant position in formal logic, including mathematical, philosophical and computational logic. But, the appearance of logics alternative to classical logic is one of the landmarks in the history of logic over the last century.

These logics are usually called *non-classical logics*. One amazing aspect of the development of non-classical logics is recognized in the fact that they can overcome some limitations of classical logic, which should be considered in applications to human knowledge.

In fact, non-classical logics can be classified as two classes. One class is a *rival* to classical logic, including *intuitionistic logic* and *many-valued logic*. The logics in this class deny some principles of classical logic.

For example, intuitionistic logic does not adopt the law of excluded middle $A \vee \neg A$ as an axiom. As a consequence, intuitionistic logic can support constructive reasoning. Therefore, intuitionistic logic is suited to mathematics and computer science.

Many-valued logic denies the two-valuedness of classical logic, allowing several truth-values. Thus, fuzzy logic belongs to a family of many-valued logics. Consequently, many-valued logic can deal with the features of information like degree and vagueness.

The second class is an *extension* of classical logic. The logics in this class thus expand classical logic with some logical machineries. One of the most notable logics in the class is *modal logic*, which extends classical logic with modal operators to formalize the notions of necessity and possibility.

As is well known, if we read modal operators differently, several *intensional logics* can be obtained. For instance, *temporal logic* is considered as a logic of time, *epistemic logic* as a logic of knowledge, *deontic logic* as a logic of obligation, and so on.

J.M. Abe et al., *Introduction to Annotated Logics*,
Intelligent Systems Reference Library 88, DOI 10.1007/978-3-319-17912-4_1

Recently, there has been a growing interest in the use of non-classical logics in several areas like mathematics, philosophy, linguistics, computer science and Artificial Intelligence (AI). This is because classical logic has some limitations in such applications. Consequently, we need to use non-classical logics. As mentioned above, it can be pointed out that classical logic is not enough to formalize the problems in several areas.

If we want to use *some* logic to model human common-sense reasoning, we must properly deal with incomplete and inconsistent information with theoretical backgrounds. This is because we generally perform daily reasoning based on information which is incomplete and (or) inconsistent. It is thus desirable to develop a logic which meets the requirement.

A lot work has been done to study foundations for such logics. However, we do not seem to have a conclusive formalism. This fact reveals that work to seek such logics is a difficult task. We believe that attractive logics in this context are undoubtedly *annotated logics*, which were originally proposed by Subrahmanian [149] in 1987 in the context of logic programming.

This book is an introduction to annotated logics which are considered as systems of non-classical logics. Although the starting point of annotated logics is computational, they have the merit of formal investigation. We believe that annotated logic is one of the *ideal* paraconsistent logics. For this reason, we aim at focusing on the foundations for annotated logics, but we also show their applications to computer science.

In general, annotated logics are a kind of paraconsistent, paracomplete and non-alethic logic. In what follows, we give some conventions and definitions used in this book.

Let T be a theory whose underlying logic is L. T is called *inconsistent* when it contains theorems of the form A and $\neg A$ (the negation of A), i.e.,

$$T \vdash_L A \text{ and } T \vdash_L \neg A$$

where $\vdash_L$ denotes the provability relation in L. If T is not inconsistent, it is called *consistent*.

T is said to be *trivial*, if all formulas of the language are also theorems of T. Otherwise, T is called *non-trivial*. Then, for trivial theory T, $T \vdash_L B$ for any formula B. Note that trivial theory is not interesting since every formula is provable.

If L is classical logic (or one of several others, such as intuitionistic logic), the notions of inconsistency and triviality agree in the sense that T is inconsistent iff T is trivial. So, in trivial theories the extensions of the concepts of formula and theorem coincide.

A *paraconsistent logic* is a logic that can be used as the basis for inconsistent but non-trivial theories. In this regard, paraconsistent theories do not satisfy the *principle of non-contradiction*, i.e., $\neg(A \wedge \neg A)$.

Similarly, we can define the notion of paracomplete theory, namely T is called *paracomplete* when neither A nor $\neg A$ is a theorem. In other words,

$$T \nvdash_L A \text{ and } T \nvdash_L \neg A$$

hold in paracomplete theory. If T is not paracomplete, T is *complete*, i.e.,

$$T \vdash_L A \text{ or } T \vdash_L \neg A$$

holds. A *paracomplete logic* is a logic for paracomplete theory, in which the *principle of excluded middle*, i.e., $A \vee \neg A$ fails. In this sense, intuitionistic logic is one of the paracomplete logics.

Finally, a logic which is simultaneously paraconsistent and paracomplete is called *non-alethic logic*. Classical logic is a consistent and complete logic, and this fact may be problematic in applications.

1.2 History

Here, we briefly review the history of annotated logics. Before presenting it, we need to look at the history of paraconsistent logics. While paraconsistent logics have recently proved attracted to many people, they have a longer history than classical logic. For instance, Aristotle developed a logical theory that can be interpreted to be paraconsistent. But, paraconsistent logics in the modern sense were formally devised in the 1950s.

In 1910, the Russian logician Nikolaj A. Vasil'ev (1880–1940) and the Polish logician Jan Łukasiewicz (1878–1956) independently glimpsed the possibility of developing paraconsistent logics. Vasil'ev's *imaginary logic* can be seen as a paraconsistent reformulation of Aristotle's *syllogistic*; see Vasil'ev [156].

It was here pointed out that Łukasiewicz's *three-valued logic* is a forerunner of the many-valued approach to paraconsistency, although he did not explicitly discuss paraconsistency; see Łukasiewicz [116].

However, we believe that the history of paraconsistent logic started in 1948. Stanislaw Jaśkowski (1896–1965) proposed a paraconsistent propositional logic, now called *discursive logic* (or discussive logic) in 1948; see Jaśkowski [95, 96]. Discursive logic is based on modal logic, and it is classified as the modal approach to paraconsistency.

Independently, some years later, the Brazilian logician Newton C.A. da Costa (1929-) constructed for the first time hierarchies of paraconsistent propositional calculi $C_i (1 \leq i \leq \omega)$ and its first-order and higher-order extensions; see da Costa [61]. da Costa's logics are called the *C-system*, which is based on the non-standard interpretation of negation which is dual to intuitionistic negation.

A different route to paraconsistent logic may be found in the so-called *relevance logic* (or relevant logic), which was originally developed by Anderson and Belnap in the 1960s; see Anderson and Belnap [34] and Anderson et al. [35]. Anderson and Belnap's approach addresses a correct interpretation of implication $A \rightarrow B$, in which A and B should have some connection. Its semantic interpretation raises the issues of paraconsistency, and some (not all) relevance logics are in fact paraconsistent.

The above three approaches constitute the major approaches to paraconsistent logics. This is true now, but a number of other paraconsistent logics have been proposed in the literature. Although the starting point of paraconsistent logics is philosophical, they are attractive to those who working on areas like logic, mathematics and computer science.

The important problems handled by paraconsistent logics include the paradoxes of set theory, the semantic paradoxes, and some issues in dialectics. These problems are central to philosophy and philosophical logic. However, paraconsistent logics have later found interesting applications in AI, in particular, expert systems, belief, and knowledge, since the 1980s; see da Costa and Subrahmanian [65].

Annotated logics were introduced by Subrahmanian to provide a foundation for paraconsistent logic programming; see Subrahmanian [149] and Blair and Subrahmanian [53]. Paraconsistent logic programming can be seen as an extension of logic programming based on classical logic. Thus, the work can be considered as an expansion of that of Kowalski [105, 106].

In 1989, Kifer and Lozinskii proposed a logic for reasoning with inconsistency, which is related to annotated logics; see Kifer and Lozinskii [98, 99]. In the same year, Kifer and Subrahmanian extended annotated logics by introducing *generalized annotated logics* in the context of logic programming; see Kifer and Subrahmanian [100]. In 1990, a resolution-style automatic theorem-proving method for annotated logics was implemented; see da Costa et al. [64].

Of course, annotated logics were developed as a foundation for paraconsistent logic programming, but they have interesting features to be examined by logicians. Formally, annotated logics are non-alethic in the sense of the above terminology. From the viewpoint of paraconsistent logicians, annotated logics were regarded as new systems.

In 1991, da Costa and others started to study annotated logics from a foundational point of view; see da Costa et al. [62, 66]. In these works, propositional and predicate annotated logics were formally investigated by presenting axiomatization, semantics and completeness results, and some applications of annotated logics were briefly surveyed.

In 1992, Jair Minoro Abe wrote a Ph.D. thesis on the foundations of annotated logics under Prof. Newton C. A. da Costa at University of Sao Paulo; see Abe [1]. Abe proposed annotated modal logics which extend annotated logics with modal operator in Abe [2]; also see Akama and Abe [23].

Some formal results including decidability annotated logics were presented in Abe and Akama [8]. Abe and Akama also investigated predicate annotated logics by the method of ultraproducts in Abe and Akama [7]. Abe [4] studied an algebraic study of annotated logics.

Later, Abe, Akama and Nakamatsu jointly continued to study the foundations and applications of annotated logics. In this book, the details of the foundations for annotated logics will be shown from various viewpoints. We will also present some intriguing applications of annotated logics. A comprehensive exposition of the applications of annotated logics to AI in the 1990s may be found in Abe [3]. The reader will see important progresses on annotated logics in this book.

Chapter 2
Propositional Annotated Logics $P\tau$

Abstract This chapter introduces the propositional annotated logics $P\tau$. We present a Hilbert style axiomatization of $P\tau$ and their semantics. We show some formal results including completeness.

2.1 Language

Here, we start by presenting the *language* of the propositional annotated logics $P\tau$ (also denoted $P\mathcal{T}$) following da Costa et al. [66]; also see Abe [1]. We denote by L the language of $P\tau$.

Annotated logics are based on some arbitrary fixed finite lattice called a *lattice of truth-values* denoted by $\tau = \langle | \tau |, \leq, \sim \rangle$, which is the complete lattice with the ordering $\leq$ and the operator $\sim : | \tau | \rightarrow | \tau |$.

Here, $\sim$ gives the "meaning" of atomic-level negation of $P\tau$. We also assume that $\top$ is the top element and $\bot$ is the bottom element, respectively. In addition, we use two lattice-theoretic operations: $\vee$ for the least upper bound and $\wedge$ for the greatest lower bound.[1]

Definition 2.1 (*Symbols*) The symbols of $P\tau$ are defined as follows:

1. Propositional symbols: $p, q, \ldots$ (possibly with subscript)
2. Annotated constants: $\mu, \lambda, \ldots \in | \tau |$
3. Logical connectives: $\wedge$ (conjunction), $\vee$ (disjunction), $\rightarrow$ (implication), and $\neg$ (negation)
4. Parentheses: (and).

Definition 2.2 (*Formulas*) Formulas are defined as follows:

1. If p is a propositional symbol and $\mu \in | \tau |$ is an annotated constant, then p_μ is a formula called an *annotated atom*.
2. If F is a formula, then $\neg F$ is a formula.

[1] We employ the same symbols for lattice-theoretical operations as the corresponding logical connectives.

J.M. Abe et al., *Introduction to Annotated Logics*,
Intelligent Systems Reference Library 88, DOI 10.1007/978-3-319-17912-4_2

3. If F and G are formulas, then $F \wedge G$, $F \vee G$, $F \rightarrow G$ are formulas.
4. If p is a propositional symbol and $\mu \in |\tau|$ is an annotated constant, then a formula of the form $\neg^k p_\mu$ $(k \geq 0)$ is called a *hyper-literal*. A formula which is not a hyper-literal is called a *complex formula*.

Here, some remarks are in order. The annotation is attached only at the atomic level. An annotated atom of the form p_μ can be read "it is believed that p's truth-value is at least μ". In this sense, annotated logics incorporate the feature of many-valued logics.

A hyper-literal is a special kind of formula in annotated logics. In the hyper-literal of the form $\neg^k p_\mu$, $\neg^k$ denotes the k's repetition of $\neg$. More formally, if A is an annotated atom, then $\neg^0 A$ is A, $\neg^1 A$ is $\neg A$, and $\neg^k A$ is $\neg(\neg^{k-1} A)$. The convention is also used for $\sim$.

Next, we define some abbreviations.

Definition 2.3 Let A and B be formulas. Then, we put:

$$A \leftrightarrow B =_{def} (A \rightarrow B) \wedge (B \rightarrow A)$$
$$\neg_* A =_{def} A \rightarrow (A \rightarrow A) \wedge \neg(A \rightarrow A)$$

Here, $\leftrightarrow$ is called the *equivalence* and $\neg_*$ *strong negation*, respectively.

Observe that strong negation in annotated logics behaves classically in that it has all the properties of classical negation.

2.2 Semantics

A *semantics* specifies the meaning of formulas in a logical system. The semantics for $P\tau$ can be given in various ways. Now, we describe a *model-theoretic semantics* for $P\tau$.

Let $\mathbf{P}$ be the set of propositional variables. An *interpretation* I is a function $I : \mathbf{P} \rightarrow \tau$. To each interpretation I, we associate a *valuation* $v_I : \mathbf{F} \rightarrow \mathbf{2}$, where $\mathbf{F}$ is a set of all formulas and $\mathbf{2} = \{0, 1\}$ is the set of truth-values. Henceforth, the subscript is suppressed when the context is clear.

Definition 2.4 (*Valuation*) A valuation v is defined as follows:
If p_λ is an annotated atom, then

$$v(p_\lambda) = 1 \text{ iff } I(p) \geq \lambda,$$
$$v(p_\lambda) = 0 \text{ otherwise},$$
$$v(\neg^k p_\lambda) = v(\neg^{k-1} p_{\sim\lambda}), \text{ where } k \geq 1.$$

If A and B are formulas, then

$$v(A \wedge B) = 1 \text{ iff } v(A) = v(B)) = 1,$$
$$v(A \vee B) = 0 \text{ iff } v(A) = v(B) = 0,$$
$$v(A \rightarrow B) = 0 \text{ iff } v(A) = 1 \text{ and } v(B) = 0.$$

If A is a complex formula, then

$$v(\neg A) = 1 - v(A).$$

We say that the valuation v *satisfies* the formula A if $v(A) = 1$ and that v *falsifies* A if $v(A) = 0$. For the valuation v, we can obtain the following lemmas.

Lemma 2.1 *Let p be a propositional variable and $\mu \in |\tau|$ $(k \geq 0)$, then we have:*

$$v(\neg^k p_\mu) = v(p_{\sim^k \mu}).$$

Proof Immediate from the definition of the valuation of hyper-literal.

Lemma 2.2 *Let p be a propositional variable, then we have:*

$$v(p_\perp) = 1$$

Proof Since $\perp$ is the bottom element, it can be derived by the definition of $v(p_\lambda)$.

Lemma 2.3 *For any complex formula A and B and any formula F, the valuation v satisfies the following:*

1. $v(A \leftrightarrow B) = 1$ iff $v(A) = v(B)$
2. $v((A \to A) \land \neg(A \to A)) = 0$
3. $v(\neg_* A) = 1 - v(A)$
4. $v(\neg F \leftrightarrow \neg_* F) = 1$

Proof (1): For ($\Rightarrow$), suppose $v(A \leftrightarrow B) = 1$. Then, we have:

$$v((A \to B) \land (B \to A)) = 1,$$

namely, both (a) $v(A \to B) = 1$ and (b) $v(B \to A) = 1$ hold. Now, assume that $v(A) \neq v(B)$. This means that either the case that $v(A) = 1$ and $v(B) = 0$ or the case that $v(A) = 0$ and $v(B) = 1$. In other words, $v(A \to B) = 0$ or $v(B \to A) = 0$. This contradicts the hypothesis that both (a) and (b). Thus, we have $v(A) = v(B)$.

For ($\Leftarrow$), suppose $v(A) = v(B)$. Then, there are two cases (a) $v(A) = v(B) = 1$ and (b) $v(A) = v(B) = 0$. For (a), by definition, we have $v(A \to B) = v(B \to A) = 1$. Then, we have $v((A \to B) \land (B \to A)) = 1$, i.e., $v(A \leftrightarrow B) = 1$. For (b), by definition, we have $v(A \to B) = v(B \to A) = 1$. Thus, $v((A \to B) \land (B \to A)) = v(A \leftrightarrow B) = 1$ follows.
(2): We have that $v(A \to A) \neq v(\neg(A \to A))$, because $v(A \to A) = 1 - v(\neg(A \to A))$. This leads to the conclusion that $v((A \to A) \land \neg(A \to A)) = 0$.
(3): We have two cases (a) $v(A) = 0$ and (b) $v(A) = 1$. For (a), $v(A \to (A \to A) \land \neg(A \to A)) = 1$. By the definition of strong negation, $v(\neg_* A) = 1$. This implies that $v(\neg_* A) = 1 - v(A)$. For (b), by (2), $v((A \to A) \land \neg(A \to A)) = 0$ holds. Then, from the valuation of implication, we have that $v(A \to (A \to A) \land \neg(A \to A)) = 0$, i.e., $v(\neg_* A) = 0$. We can thus conclude that $v(\neg_* A) = 1 - v(A)$.
(4): Suppose that $v(\neg F) \neq v(\neg_* F)$. Here, $v(F) = 1 - v(\neg_* F)$ from (3). The valuation of $\neg$ gives rise to the fact that $v(F) = 1 - v(\neg F)$. From these facts, $v(\neg F) = v(\neg_* F)$ follows. But, it contradicts the assumption. Therefore, $v(\neg F \leftrightarrow \neg_* F) = 1$ by (1).

We here define the notion of semantic consequence relation denoted by $\models$. Let Γ be a set of formulas and F be a formula. Then, F is a *semantic consequence* of Γ, written $\Gamma \models F$, iff for every v such that $v(A) = 1$ for each $A \in \Gamma$, it is the case that $v(F) = 1$. If $v(A) = 1$ for each $A \in \Gamma$, then v is called a *model* of Γ. If Γ is empty, then $\Gamma \models F$ is simply written as $\models F$ to mean that F is *valid*.

Lemma 2.4 *Let p be a propositional variable and $\mu, \lambda \in \mid \tau \mid$. Then, we have:*

1. $\models p_\bot$
2. $\models p_\mu \to p_\lambda,\ \mu \geq \lambda$
3. $\models \neg^k p_\mu \leftrightarrow p_{\sim^k \mu},\ k \geq 0$

Proof (1): By the definition of v, we have $I(p) \geq \bot$ for any interpretation I. Therefore, $\models p_\bot$ holds.
(2): Suppose that there exists an interpretation such that $\not\models p_\mu \to p_\lambda$. This implies that $\models p_\mu$ and $\not\models p_\lambda$. So, $I(p) \geq \mu$ and $I(p) \not\geq \lambda$. This contradicts the assumption. Thus, we have that $\models p_\mu \to p_\lambda$, if $\mu \geq \lambda$.
(3): Immediate from Lemma 2.1.

The consequence relation $\models$ satisfies the next property.

Lemma 2.5 *Let A, B be formulas. Then, if $\models A$ and $\models A \to B$ then $\models B$.*

Proof Suppose $v(A) = 1$ and $v(A \to B) = 1$ but $v(B) = 0$. By the definition of $v(A \to B)$, we have $v(A \to B) = 0$ contradicting the assumption.

Lemma 2.6 *Let F be a formula, p a propositional variable, and $(\mu_i)_{i \in J}$ be an annotated constant, where J is an indexed set. Then, if $\models F \to p_\mu$, then $F \to p_{\mu_i}$, where $\mu = \bigvee \mu_i$.*

Proof Suppose that we have the valuation such that $v(F \to p_{\mu_i}) = 1$ and $v(F \to p_\mu) = 0$. Then, $v(F) = 1, v(p_{\mu_i}) = 1$ and $v(\mu) = 0$ hold. The third condition implies that $I(p) \not\geq \mu$. In other words, we have that $I(p) \not\geq \mu_j$ with $j \in J$. But, by the second condition, $I(p) \geq \mu_j$ holds. Since $\mu = \bigvee \mu_j$ holds, we have that $v(F \to p_\mu) = 1$. A contradiction.

As a corollary to Lemma 2.6, we can obtain the following lemma.

Lemma 2.7 $\models p_{\lambda_1} \wedge p_{\lambda_2} \wedge \cdots \wedge p_{\lambda_m} \to p_\lambda$, *where* $\lambda = \bigvee_{i=1}^{m} \lambda_i$.

Next, we discuss some results related to paraconsistency and paracompleteness.

Definition 2.5 (*Complementary property*) A truth-value $\mu \in \tau$ has the *complementary property* if there is a λ such that $\lambda \leq \mu$ and $\sim \lambda \leq \mu$. A set $\tau' \subseteq \tau$ has the *complementary property* iff there is some $\mu \in \tau'$ such that μ has the complementary property.

Definition 2.6 (*Range*) Suppose I is an interpretation of the language L. The *range* of I, denoted $range(I)$, is defined to be $range(I) = \{\mu \mid (\exists A \in B_L)I(A) = \mu\}$, where B_L denotes the set of all ground atoms in L.

For $P\tau$, ground atoms correspond to propositional variables. If the range of the interpretation I satisfies the complementary property, then the following theorem can be established.

Theorem 2.1 *Let I be an interpretation such that $range(I)$ has the complementary property. Then, there is a propositional variable p and $\mu \in |\tau|$ such that*

$$v(p_\mu) = v(\neg p_\mu) = 1.$$

Proof Since $range(I)$ has the complementary property, there is a propositional variable p and a $\delta \in \tau$, satisfying (1) $I(p) = \delta$ and (2) there is a $\gamma \in \tau$ such that $\gamma \leq \delta$ and $\sim\gamma \leq \delta$. By (1), $I(p) \geq \delta$ holds. Thus, we have that $v(p_\gamma) = 1$. Similarly, we have that $v(\neg p_\gamma) = 1$ by (2). From both, we can reach the theorem by Definition 2.5.

Theorem 2.1 states that there is a case in which for some propositional variable it is both true and false, i.e., inconsistent. The fact is closely tied with the notion of paraconsistency.

Definition 2.7 ($\neg$-*inconsistency*) We say that an interpretation I is $\neg$-*inconsistent* iff there is a propositional variable p and an annotated constant $\mu \in |\tau|$ such that $v(p_\mu) = v(\neg p_\mu) = 1$.

Therefore, $\neg$-inconsistency means that both A and $\neg A$ are simultaneously true for some atomic A. Below, we formally define the concepts of non-triviality, paraconsistency and paracompleteness.

Definition 2.8 (*Non-triviality*) We say that an interpretation I is *non-trivial* iff there is a propositional variable p and an annotated constant $\mu \in |\tau|$ such that $v(p_\mu) = 0$.

By Definition 2.8, we mean that not every atom is valid if an interpretation is non-trivial.

Definition 2.9 (*Paraconsistency*) We say that an interpretation I is *paraconsistent* iff it is both $\neg$-inconsistent and non-trivial. $P\tau$ is called *paraconsistent* iff there is an interpretation of I of $P\tau$ such that I is paraconsistent.

Definition 2.9 allows the case in which both A and $\neg A$ are true, but some formula B is false in some paraconsistent interpretation I.

Definition 2.10 (*Paracompleteness*) We say that an interpretation I is *paracomplete* iff there is a propositional variable p and an annotated constant $\lambda \in |\tau|$ such that $v(p_\lambda) = v(\neg p_\lambda) = 0$. $P\tau$ is called *paracomplete* iff there is an interpretation I of $P\tau$ such that I is paracomplete.

From Definition 2.10, we can see that in the paracomplete interpretation I, both A and $\neg A$ are false. We say that $P\tau$ is *non-alethic* iff it is both paraconsistent and paracomplete. Intuitively speaking, paraconsistent logic can deal with inconsistent information and paracomplete logic can handle incomplete information.

This means that non-alethic logics like annotated logics can serve as logics for expressing both inconsistent and incomplete information. This is one of the starting points of our study of annotated logics.

As the following Theorems 2.2 and 2.3 indicate, paraconsistency and paracompleteness in $P\tau$ depend on the cardinality of τ.

Theorem 2.2 *$P\tau$ is paraconsistent iff $card(\tau) \geq 2$, where $card(\tau)$ denotes the cardinality (cardinal number) of the set τ.*

Proof For ($\Rightarrow$), let I be a paraconsistent interpretation. Then, there are propositional variables p, q and annotated constants μ, λ such that (1) $v(p_\mu) = v(\neg p_\mu) = 1$ and (2) $v(q_\lambda) = 0$. Since τ is a complete lattice, $card(\mu) \geq 1$. Now, assume that $card(\mu) = 1$, namely τ has one element μ, and $\mu = \bot = \top$. This means that $\bot$ and $\top$ agree in the lattice. Here, the only possible interpretation is to assign $\mu = \bot = \top$ to all propositional variables. It follows that for any propositional variable r, $v(r_\top) = v(r_\bot) = 1$. But, it contradicts the condition (2).

For ($\Leftarrow$), suppose $card(\tau) \geq 2$, with $\bot \neq \top$. Here, we can define the interpretation I such that $I(p) = \top$ and $I(q) = \bot$. Then, $v(p_\top) = 1$ follows. As $\sim\top \leq \top$, we have that $v(p_{\sim\top}) = v(\neg p_\top) = 1$. Since $\bot \leq \top$, we have that $v(q_\top) = 0$. Consequently, I is paraconsistent, and we can conclude that $P\tau$ is paraconsistent.

Theorem 2.3 *$P\tau$ is paracomplete iff $card(\tau) \geq 2$.*

Proof For ($\Rightarrow$), let I be a paracomplete interpretation. Then, there is a propositional variable p and an annotated constant μ such that $v(p_\mu) = v(\neg p_\mu) = 0$. Here, we assume that $card(\tau) = 1$ and set $\mu = \bot = \top$. Thus, $v(p_\bot) = v(p_\top) = 0$ holds. But, the only possible interpretation is $v(p_\bot) = v(p_\top) = 1$ for any propositional variable p. This is a contradiction.

For ($\Leftarrow$), suppose $card(\tau) \geq 2$, with $\bot \neq \top$. Now, assume that $I(p) = \bot$. Then, $v(p_\top) = 0$ since $I(p) = \bot \ngeq \top$. Now, we define the negation operator $\sim$ satisfying $\sim\top = \top$. Therefore, $v(\neg p_\top) = v(p_{\sim\top}) = v(p_\top) = 0$. As a consequence, I is shown to be a paracomplete interpretation. In other words, $P\tau$ is paracomplete.

The above two theorems imply that to formalize a non-alethic logic based on annotated logics we need at least both the top and bottom elements of truth-values. The simplest lattice of truth-values is *FOUR* in Belnap [49, 50], which is shown in Fig. 2.1.

Definition 2.11 (*Theory*) Given an interpretation I, we can define the theory $Th(I)$ associated with I to be a set:

$$Th(I) = Cn(\{p_\mu \mid p \in \mathbf{P} \text{ and } I(p) \geq \mu\}).$$

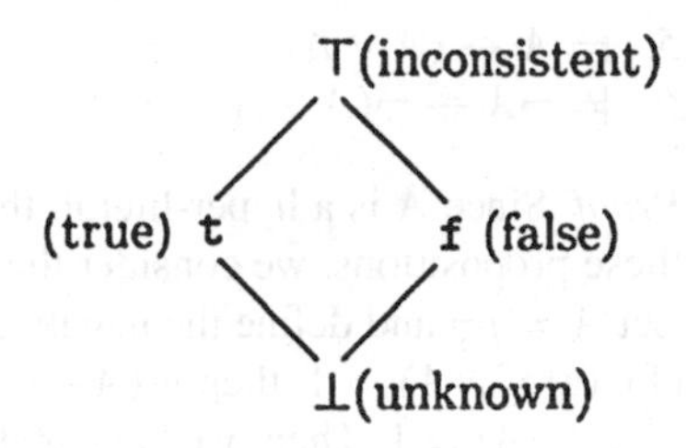

Fig. 2.1 The lattice *FOUR*

Here, Cn is the semantic consequence relation, i.e.,

$$Cn(\Gamma) = \{F \mid F \in \mathbf{F} \text{ and } \Gamma \models F\}.$$

Here, Γ is a set of formulas.

$Th(I)$ can be extended for any set of formulas.

Theorem 2.4 *An interpretation I is $\neg$-inconsistent iff $Th(\Gamma)$ is $\neg$-inconsistent.*

Proof For ($\Rightarrow$), suppose I is $\neg$-inconsistent. Then, there is an interpretation I and a hyper-literal p_μ such that $v(\neg p_\mu) = 1$. By the definition of $Th(I)$, we have that $p_\mu, \neg p_\mu \in Th(I)$. Next, consider a complex formula A for $\neg$-inconsistent interpretation. $v(A) = v(\neg A) = 1$ holds. Because $v(\neg A) = 1 - v(A)$, we have that $v(A) \neq v(\neg A)$. But it contradicts the assumption. Therefore, $Th(I)$ is $\neg$-inconsistent.

For ($\Leftarrow$), suppose $Th(I)$ is $\neg$-inconsistent. Then, $p_\mu, \neg p_\mu \in Th(I)$. It follows that $v(p_\mu) = v(\neg p_\mu) = 1$, concluding that I is $\neg$-inconsistent. Next, for complex formula A, $A, \neg A \in Th(I)$ holds. Then, we can $v(A) = 1$ and $v(\neg A)$ follows since $v(\neg A) = 1$. From these facts, I is $\neg$-inconsistent.

Theorem 2.5 *An interpretation I is paraconsistent iff $Th(I)$ is paraconsistent.*

Proof For ($\Rightarrow$), let I be a paraconsistent interpretation. Then, for some hyper-literals p_μ, q_μ, we have that $v(p_\mu = v(\neg p_\mu) = 1$ and $v(q_\mu) = 0$. Then, $p_\mu, \neg p_\mu \in Th(I)$, but $q_\mu \notin Th(I)$. For complex formulas A, B, we have that $v(A) = v(\neg A) = 1$ and $v(B) = 0$. From this, $A, \neg A \in Th(I)$, but $B \notin Th(I)$. This means that $Th(I)$ is paraconsistent.

For ($\Leftarrow$), $p_\mu, \neg p_\mu \in Th(I)$ but $q_\mu \notin Th(I)$ for some hyper-literals p_μ, q_μ. Then, $v(p_\mu) = v(\neg p_\mu) = 1$ and $v(q_\mu) = 0$ follows. This shows that I is paraconsistent.

The next lemma states that the replacement of equivalent formulas within the scope of $\neg$ does not hold in $P\tau$ as in other paraconsistent logics.

Lemma 2.8 *Let A be any hyper-literal. Then, we have:*

1. $\models A \leftrightarrow ((A \rightarrow A) \rightarrow A)$
2. $\not\models \neg A \leftrightarrow \neg(((A \rightarrow A) \rightarrow A))$
3. $\models A \leftrightarrow (A \wedge A)$
4. $\not\models \neg A \leftrightarrow \neg(A \wedge A)$

5. $\models A \leftrightarrow (A \vee A)$
6. $\not\models \neg A \leftrightarrow \neg(A \vee A)$

Proof Since A is a hyper-literal, then A is of the form p_μ, and $\mu \in |\tau|$. To prove these propositions, we consider the cases that $v(A) = 1$ and the case that $v(A) = 0$. Let $A = p_\top$ and define the negation operator $\sim$ satisfying $\sim\top = \top$.
(1): (a) If $v(A) = 1$, then $v((A \to A) \to A) = 1$ as required. (b) if $v(A) = 0$, then $v(A \to A) = 1$. Then, we have that $v((A \to A) \to A) = 0 = v(A)$.
(2) (a) If $v(p_\top) = 1$, then $v(\neg p_\top) = v(p_{\sim\top}) = v(p_\top) = 1$. Here, we have that $v(\neg((p_\top \to p_\top) \to p_\top)) = 1 - v((p_\top \to p_\top) \to p_\top) = 0 \neq v(p_\top)$. (b) If $v(p_\top) = 0$, then $v(\neg p_\top) = 0$. Here, we have that $v(\neg((p_\top \to p_\top) \to p_\top)) = 1 \neq v(p_\top)$.
(3) (a) If $v(A) = 1$, then $v(A \wedge A) = 1$. (b) If $v(A) = 0$, then $v(A \wedge A) = 0$.
(4) (a) If $v(p_\top) = 1$, then $v(\neg p_\top) = 1$. We can see that $v(\neg(p_\top \wedge p_\top)) = 1 - v(p_\top \wedge p_\top) = 0 \neq v(p_\top)$. (b) If $v(p_\top) = 0$, then $v(\neg p_\top) = 0$. Here, $v(\neg(p_\top \wedge p_\top)) = 1 - v(p_\top \wedge p_\top) = 1 \neq v(p_\top)$, as required.

As obvious from the above proofs, (1), (3) and (5) hold for any formula A. But, (2), (4) and (6) cannot be generalized for any A.

By the next theorem, we can find the connection of $P\tau$ and the positive fragment of classical propositional logic C.

Theorem 2.6 *If $F_1, \ldots, F_n$ are complex formulas and $K(A_1, \ldots, A_n)$ is a tautology of C, where $A_1, \ldots, A_n$ are the sole propositional variable occurring in the tautology, then $K(F_1, \ldots, F_n)$ is valid in $P\tau$. Here, $K(F_1, \ldots, F_n)$ is obtained by replacing each occurrence of A_i, $1 \leq i \leq n$, in K by F_i.*

Proof Proved by induction on n. For example, consider the formulas

$$K(p, q) = (p \wedge q) \to (q \wedge p)$$

which is a well-known tautology of C. Let $F_1 = F$, $F_2 = \neg G$ be complex formulas. By definition,

$$K(F, \neg G) = (F \wedge \neg G) \to (\neg G \wedge F)$$

is obtained. It suffices to show that $K(F, \neg G)$ is valid in $P\tau$. In other words, $v(K(F, \neg G)) = 1$ for any v. Suppose that $v(K(F, \neg G)) = 1$. This is equivalent to the following:

$v(F \wedge \neg G) = 0$ or $v(\neg G \wedge F) = 1$
iff ($v(F) = 0$ or $v(\neg G) = 0$) or ($v(\neg G) = 1$ and $v(F) = 1$)
iff ($v(F) = 0$ or $v(G) = 1$) or ($v(G) = 0$ and $v(F) = 1$)

It is easy to check that the last clause is satisfied by any v. Thus, $\models K(F, \neg G)$, that is, $K(F, \neg G)$ is valid in $P\tau$.

Next, we consider the properties of strong negation $\neg_*$.

Theorem 2.7 *Let A, B be any formulas. Then,*

1. $\models (A \rightarrow B) \rightarrow ((A \rightarrow \neg_* B) \rightarrow \neg_* A)$
2. $\models A \rightarrow (\neg_* A \rightarrow B)$
3. $\models A \vee \neg_* A$

Proof For (1), assume that there is a valuation v such that $v(A \rightarrow B) = 1$ and $v((A \rightarrow \neg_* B) \rightarrow \neg_* A) = 0$. From the latter, we have that $v(A \rightarrow \neg_* B) = 1$ and $v(\neg_* A) = 0$. Thus, $v(A) = 1$ because $v(A) = 1 - v(\neg_* A)$. From the former, we have that $v(A \rightarrow B) = 1$. So $v(A) = 0$ or $v(B) = 1$ holds. But, the claim that $v(A) = 1$ enables us to infer that $v(B) = 1$. As a consequence, we have:

$$v(A) = 1 \text{ and } v(B) = 1$$

Here, the latter also ensures that $v(A \rightarrow \neg_* B) = 1$. However, since $v(A) = 1$, we have to obtain $v(\neg_* B) = 1$, i.e. $v(B) = 0$. This induces a contradiction.

For (2), assume that we have a v satisfying that $v(A) = 1$ and $v(\neg_* A \rightarrow B) = 0$. The latter gives rise to the condition that $v(\neg_* A) = 1$ and $v(B) = 0$. Here, we have that $v(A) = 0$ from $v(\neg_* A) = 1$. However, this is impossible.

For (3), assume the existence of v such that $v(A \vee \neg_* A) = 0$. This implies that $v(A) = v(\neg_* A) = 0$. However, this is impossible.

Theorem 2.7 tells us that strong negation has all the basic properties of classical negation. Namely, (1) is a principle of *reductio ad abusurdum*, (2) is the related principle of the law of non-contradiction, and (3) is the law of excluded middle. Note that $\neg$ does not satisfy these properties. It is also noticed that for any complex formula $A \models \neg A \leftrightarrow \neg_* A$ but that for any hyper-literal $Q \not\models \neg Q \leftrightarrow \neg_* Q$.

From these observations, $P\tau$ is a paraconsistent and paracomplete logic, but adding strong negation enables us to perform classical reasoning.

2.3 Axiomatization

In this section, we provide an axiomatization of $P\tau$ in the Hilbert style. There are many ways to axiomatize a logical system, one of which is the *Hilbert system*. We discuss other proof systems in Chap. 4.

A Hilbert system can be defined by the set of *axioms* and *rules of inference*. Here, an axiom is a formula to be postulated as valid, and rules of inference specify how to prove a formula.

We are now ready to give a Hilbert style axiomatization of $P\tau$, called $\mathscr{A}\tau$. Let A, B, C be arbitrary formulas, F, G be complex formulas, p be a propositional variable, and λ, μ, λ_i be annotated constant. Then, the postulates are as follows (cf. Abe [1]):

Postulates for $\mathscr{A}\tau$

($\rightarrow_1$) $(A \rightarrow (B \rightarrow A)$
($\rightarrow_2$) $(A \rightarrow (B \rightarrow C)) \rightarrow ((A \rightarrow B) \rightarrow (A \rightarrow C))$

($\rightarrow_3$) $((A \rightarrow B) \rightarrow A) \rightarrow A$
($\rightarrow_4$) $A, A \rightarrow B/B$
($\wedge_1$) $(A \wedge B) \rightarrow A$
($\wedge_2$) $(A \wedge B) \rightarrow B$
($\wedge_3$) $A \rightarrow (B \rightarrow (A \wedge B))$
($\vee_1$) $A \rightarrow (A \vee B)$
($\vee_2$) $B \rightarrow (A \vee B)$
($\vee_3$) $(A \rightarrow C) \rightarrow ((B \rightarrow C) \rightarrow ((A \vee B) \rightarrow C))$
($\neg_1$) $(F \rightarrow G) \rightarrow ((F \rightarrow \neg G) \rightarrow \neg F)$
($\neg_2$) $F \rightarrow (\neg F \rightarrow A)$
($\neg_3$) $F \vee \neg F$
(τ_1) $p_\perp$
(τ_2) $\neg^k p_\lambda \leftrightarrow \neg^{k-1} p_{\sim\lambda}$
(τ_3) $p_\lambda \rightarrow p_\mu$, where $\lambda \geq \mu$
(τ_4) $p_{\lambda_1} \wedge p_{\lambda_2} \wedge \cdots \wedge p_{\lambda_m} \rightarrow p_\lambda$, where $\lambda = \bigvee_{i=1}^{m} \lambda_i$

Here, except ($\rightarrow_4$), these postulates are axioms. ($\rightarrow_4$) is a rule of inferences called *modus ponens* (MP).

In da Costa et al. [66], a different axiomatization is given, but it is essentially the same as ours. There, the postulates for implication are different. Namely, although ($\rightarrow_1$) and ($\rightarrow_3$) are the same (although the naming differs), the remaining axiom is:

$$(A \rightarrow B) \rightarrow ((A \rightarrow (B \rightarrow C)) \rightarrow (A \rightarrow C))$$

It is well known that there are many ways to axiomatize the implicational fragment of classical logic C. In the absence of negation, we need the so-called *Pierce's law* ($\rightarrow_3$) for C.

In ($\neg_1$), ($\neg_2$), ($\neg_3$), F and G are complex formulas. In general, without this restriction on F and G, these are not sound rules due to the fact that they are not admitted in annotated logics.

da Costa et al. [66] fuses (τ_1) and (τ_2) as the single axiom in conjunctive form. But, we separate it into two axioms for our purposes. Also there is a difference in the final axiom. They present it for infinite lattices as

$$A \rightarrow p_{\lambda_j} \text{ for every } j \in J, \text{ then } A \rightarrow p_\lambda, \text{ where } \lambda = \bigvee_{j \in J} \lambda_j.$$

If τ is a finite lattice, this is equivalent to the form of (τ_2).

As usual, we can define a *syntactic consequence relation* in $P\tau$. Let Γ be a set of formulas and G be a formula. Then, G is a syntactic consequence of Γ, written $\Gamma \vdash G$, iff there is a finite sequence of formulas $F_1, F_2, \ldots, F_n$, where F_i belongs to Γ, or F_i is an axiom ($1 \leq i \leq n$), or F_j is an immediate consequence of the previous two formulas by ($\rightarrow_4$). This definition can extend for the transfinite case in which n is an ordinal number. If $\Gamma = \emptyset$, i.e. $\vdash G$, G is a *theorem* of $P\tau$.

Let Γ, Δ be sets of formulas and A, B be formulas. Then, the consequence relation $\vdash$ satisfies the following conditions.

1. if $\Gamma \vdash A$ and $\Gamma \subset \Delta$ then $\Delta \vdash A$.
2. if $\Gamma \vdash A$ and $\Delta, A \vdash B$ then $\Gamma, \Delta \vdash B$.
3. if $\Gamma \vdash A$, then there is a finite subset $\Delta \subset \Gamma$ such that $\Delta \vdash A$.

In the Hilbert system above, the so-called *deduction theorem* holds.

Theorem 2.8 (Deduction theorem) *Let Γ be a set of formulas and A, B be formulas. Then, we have:*

$$\Gamma, A \vdash B \Rightarrow \Gamma \vdash A \to B.$$

Proof See Kleene [103].

The following theorem shows some theorems related to strong negation.

Theorem 2.9 *Let A and B be any formula. Then,*

1. $\vdash A \vee \neg_* A$
2. $\vdash A \to (\neg_* A \to B)$
3. $\vdash (A \to B) \to ((A \to \neg_* B) \to \neg_* A)$

Proof For (1), we use the theorem $A \vee (A \to B)$ of classical propositional logic. If we set $B = (A \to A) \wedge \neg(A \to A)$, then we obtain the following:

$$\vdash A \vee \neg_* A$$

by the definition of strong negation, as required.

For (2), the following two hold:

(a) $(A \to A) \wedge \neg(A \to A) \vdash B$
(b) $A, A \to (A \to A) \wedge \neg(A \to A) \vdash (A \to A) \wedge \neg(A \to A)$

By the property of $\vdash$, we have (c).

(c) $A, A \to (A \to A) \wedge \neg(A \to A) \vdash B$

By applying the deduction theorem to (c) twice, (d) is obtained.

(d) $\vdash A \to ((A \to (A \to A) \wedge \neg(A \to A)) \to B)$

By the definition of $\neg_*$, (e) follows.

(e) $\vdash A \to (\neg_* A \to B)$.

For (3), we use the theorem:

$$(A \to B) \to ((A \to (B \to C)) \to ((A \to B) \to (A \to C)).$$

Now, set $C = (A \to A) \wedge \neg(A \to A)$. Then, we have (a):

(a) $\vdash (A \to B) \to ((A \to (B \to (A \to A) \wedge \neg(A \to A)) \to (A \to (A \to A) \wedge \neg(A \to A)))$

By the definition of $\neg_*$, we can reach the following:

(b) $\vdash (A \to B) \to ((A \to \neg_* B) \to \neg_* A)$

From Theorems 2.9 and 2.10 follows.

Theorem 2.10 *For arbitrary formulas A and B, the following hold:*

1. $\vdash \neg_*(A \wedge \neg_* A)$
2. $\vdash A \leftrightarrow \neg_*\neg_* A$
3. $\vdash (A \wedge B) \leftrightarrow \neg_*(\neg_* A \vee \neg_* B)$
4. $\vdash (A \to B) \leftrightarrow (\neg_* A \vee B)$
5. $\vdash (A \vee B) \leftrightarrow \neg_*(\neg_* A \wedge \neg_* B)$

Theorem 2.10 implies that by using strong negation and a logical connective other logical connectives can be defined as in classical logic. If $\tau = \{t, f\}$, with its operations appropriately defined, we can obtain classical propositional logic in which $\neg_*$ is classical negation.

2.4 Formal Results

In this section, we provide some formal results of $P\tau$ including completeness and decidability.

Lemma 2.9 *Let p be a propositional variable and $\mu, \lambda, \theta \in |\tau|$. Then, the following hold:*

1. $\vdash p_{\lambda \vee \mu} \to p_\lambda$
2. $\vdash p_{\lambda \vee \mu} \to p_\mu$
3. $\lambda \geq \mu$ *and* $\lambda \geq \theta \Rightarrow \vdash p_\lambda \to p_{\mu \vee \theta}$
4. $\vdash p_\mu \to p_{\mu \wedge \theta}$.
5. $\vdash p_\theta \to p_{\mu \wedge \theta}$.
6. $\lambda \leq \mu$ *and* $\lambda \leq \theta \Rightarrow \vdash p_{\mu \wedge \theta}$
7. $\vdash p_\mu \leftrightarrow p_{\mu \vee \mu}$, $\vdash p_\mu \leftrightarrow p_{\mu \wedge \mu}$
8. $\vdash p_{\mu \vee \lambda} \leftrightarrow p_{\lambda \vee \mu}$, $\vdash p_{\mu \wedge \lambda} \leftrightarrow p_{\lambda \wedge \mu}$
9. $\vdash p_{(\mu \vee \lambda) \vee \theta} \vee \to p_{\mu \vee (\lambda \vee \theta)}$, $\vdash p_{(\mu \wedge \lambda) \wedge \theta} \vee \to p_{\mu \wedge (\lambda \wedge \theta)}$
10. $p_{(\mu \vee \lambda) \wedge \mu} \to p_\mu$, $p_{(\mu \wedge \lambda) \vee \mu} \to p_\mu$
11. $\lambda \leq \mu \Rightarrow \vdash p_{\lambda \vee \mu} \to p_\mu$
12. $\lambda \vee \mu = \mu \Rightarrow \vdash p_\mu \to p_\lambda$
13. $\mu \geq \lambda \Rightarrow \forall \theta \in |\tau|$ $(\vdash p_{\mu \vee \theta} \to p_{\lambda \vee \theta}$ *and* $\vdash p_{\mu \wedge \theta} \to p_{\lambda \wedge \theta})$
14. $\mu \geq \lambda$ *and* $\theta \geq \varphi \Rightarrow \vdash p_{\mu \vee \theta} \to p_{\lambda \vee \varphi}$ *and* $p_{\mu \wedge \theta} \to p_{\lambda \wedge \varphi}$

15. $\vdash p_{\mu\wedge(\lambda\vee\theta)} \rightarrow p_{(\mu\wedge\lambda)\vee(\mu\wedge\theta)}$, $\vdash p_{\mu\vee(\lambda\wedge\theta)} \rightarrow p_{(\mu\vee\lambda)\wedge(\mu\vee\theta)}$
16. $\vdash p_\mu \wedge p_\lambda \leftrightarrow p_{\mu\wedge\lambda}$
17. $\vdash p_{\mu\vee\lambda} \rightarrow p_\mu \vee p_\lambda$

Proof Immediate from the properties of τ.

Example 2.1 Consider the complete lattice $\tau = N \cup \omega$, where N is the set of natural numbers. The ordering on τ is the usual ordering on ordinals, restricted to the set τ. Consider the set $\Gamma = \{p_0, p_1, p_2, \ldots\}$, where $p_\omega \notin \Gamma$. It is clear that $\Gamma \vdash p_\omega$, but an infinitary deduction is required to establish this.

Definition 2.12 $\overline{\Delta} = \{A \in \mathbf{F} \mid \Delta \vdash A\}$.

Definition 2.13 Δ is said to be *trivial* iff $\overline{\Delta} = \mathbf{F}$ (i.e., every formula in our language is a syntactic consequence of Δ); otherwise, Δ is said to be *non-trivial*. Δ is said to be *inconsistent* iff there is some formula A such that $\Delta \vdash A$ and $\Delta \vdash \neg A$; otherwise, Δ is *consistent*.

From the definition of triviality, the next theorem follows:

Theorem 2.11 *Δ is trivial iff $\Delta \vdash A \wedge \neg A$ (or $\Delta \vdash A$ and $\Delta \vdash \neg_* A$) for some formula A.*

Proof Obvious from the axiom $(\neg_2)$ and Theorem 2.9(2).

Theorem 2.12 *Let Γ be a set of formulas, A, B be any formulas, and F be any complex formula. Then, the following hold.*

1. $\Gamma \vdash A$ *and* $\Gamma \vdash A \rightarrow B \Rightarrow \Gamma \vdash B$
2. $A \wedge B \vdash A$
3. $A \wedge B \vdash B$
4. $A, B \vdash A \wedge B$
5. $A \vdash A \vee B$
6. $B \vdash A \vee B$
7. $\Gamma, A \vdash C$ *and* $\Gamma, B \vdash C \Rightarrow \Gamma, A \vee B \vdash C$
8. $\vdash F \leftrightarrow \neg_* F$
9. $\Gamma, A \vdash B$ *and* $\Gamma, A \vdash \neg_* B \Rightarrow \Gamma \vdash \neg_* A$
10. $\Gamma, A \vdash B$ *and* $\Gamma, \neg_* A \vdash B \Rightarrow \Gamma \vdash B$.

Proof We here only prove (8), (9) and (10). For (8), we first prove $\neg F \rightarrow \neg_* F$. By definition, we have $\neg_* F =_{def} F \rightarrow (F \rightarrow F) \wedge \neg(F \rightarrow F)$. Since $\neg F, F \vdash (F \rightarrow F) \wedge \neg(F \rightarrow F)$, by the deduction theorem,

(a) $\vdash \neg F \rightarrow (F \rightarrow (F \rightarrow F) \wedge \neg(F \rightarrow F))$

holds. By the definition of $(\neg_*)$, we have (b).

(b) $\vdash \neg F \to \neg_* F$

Next, we prove $\neg_* F \to \neg F$. By $(\neg_1)$, we have (a):

(a) $\vdash (F \to (F \to F) \wedge \neg(F \to F)) \to ((F \to \neg((F \to F) \wedge \neg(F \to F))) \to \neg F)$

By hypothesis, (b) holds.

(b) $\vdash F \to (F \to F) \wedge \neg(F \to F) = \neg_* F$

By $(\to_4)$ from (a) and (b), we obtain (c).

(c) $\vdash (F \to \neg((F \to F) \wedge \neg(F \to F))) \to \neg F$

By $(\to_1)$, (d) can be derived.

(d) $\vdash \neg((F \to F) \wedge \neg(F \to F)) \to (F \to \neg((F \to F) \wedge \neg(F \to F)))$

To proceed, we prove $\neg(G \wedge \neg G)$ for any complex formula G. By $(\neg_1)$, we have (e):

(e) $\vdash ((G \wedge \neg G) \to G) \to (((G \wedge \neg G) \to \neg G) \to \neg(G \wedge \neg G))$

By $(\wedge_1)$, $(\wedge_2)$, (f) and (g) hold.

(f) $\vdash (G \wedge \neg G) \to G$
(g) $\vdash (G \wedge \neg G) \to \neg G$

By $(\to_4)$ from (e) and (f), we have (h).

(h) $\vdash ((G \wedge \neg G) \to \neg G) \to \neg(G \wedge \neg G)$

By $(\to_4)$ from (g) and (h), we have (i).

(i) $\vdash \neg(G \wedge \neg G)$

Set $G = F \to F$. Then, (i) is (j).

(j) $\neg((F \to F) \wedge \neg(F \to F))$

By $(\to_4)$ from (d) and (j), (k) is derived.

(k) $F \to \neg((F \to F) \wedge \neg(F \to F))$

Applying $(\to_4)$ to (c) and (k), we obtain (l).

(l) $\neg F$

Therefore, (m) is proved.

(m) $\vdash \neg_* F \to F$

For (9), by the deduction theorem, we have (a) and (b) from the assumptions.

(a) $\Gamma \vdash A \to B$
(b) $\Gamma \vdash A \to \neg_* B$

By Theorem 2.9(3), (c) can be derived.

(c) $\vdash (A \to B) \to ((A \to \neg_* B) \to \neg_* A)$

By ($\to_4$) from (a) and (c), we obtain (d).

(d) $\Gamma \vdash (A \to \neg_* B) \to \neg_* A$

By ($\to_4$) from (b) and (d), we obtain (e).

(e) $\Gamma \vdash \neg_* A$

For (10), assume that $\Gamma, A \vdash B$ and $\Gamma, \neg_* A \vdash A$. By Theorem 2.12(7), (a) holds.

(a) $\Gamma, A \vee \neg_* A \vdash B$

By the deduction theorem, (b) is proved.

(b) $\Gamma \vdash (A \vee \neg_* A) \to B$

By Theorem 2.9, we have (c).

(c) $\vdash A \vee \neg_* A$

By Theorem 2.12(1) from (b) and (c), (d) can be derived.

(d) $\Gamma \vdash B$

Note here that the counterpart of Theorem 2.12(10) obtained by replacing the occurrence of $\neg_*$ by $\neg$ is not valid.

Now, we are in a position to prove the soundness and completeness of $P\tau$. Our proof method for completeness is based on maximal non-trivial set of formulas; see Abe [1] and Abe and Akama [8]. da Costa et al. [66] presented another proof using Zorn's Lemma.

Theorem 2.13 (Soundness) *Let Γ be a set of formulas and A be any formula. $\mathscr{A}\tau$ is a sound axiomatization of $P\tau$, i.e., if $\Gamma \vdash A$ then $\Gamma \models A$.*

Proof It is easy to prove that all the postulates of $\mathscr{A}\tau$ is valid. From the property of $\vdash$, soundness follows.

For proving the completeness theorem, we need some theorems.

Theorem 2.14 *Let Γ be a non-trivial set of formulas. Suppose that τ is finite. Then, Γ can be extended to a maximal (with respect to inclusion of sets) non-trivial set with respect to* **F**.

Proof Let Γ be a non-trivial subset of formulas of **F**. To show that Γ can be extended to a maximal non-trivial set, we construct a sequence $\Gamma_0, \Gamma_1, \ldots, \Gamma_n, \ldots$ as follows. As a vocabulary is composed by a denumerable set of symbols, the set of formulas of **F** is denumerable. Let $\Gamma_0 = \Gamma$ and inductively construct the rest of the sequence by taking $\Gamma_{i+1} = \Gamma \cup \{A_{i+1}\}$ if this set is non-trivial and otherwise by taking $\Gamma_{i+1} = \Gamma_i$.

It is easy to see that each set of the sequence $\Gamma_0, \Gamma_1, \ldots$ is non-trivial, and this is a non-decreasing sequence of sets such that $\Gamma \subseteq \Gamma_0 \subseteq \Gamma_1 \subseteq \cdots \subseteq \cdots \Gamma \subseteq \cdots$ Set Γ^* as follows:

$$\Gamma^* = \bigcup_{i=0}^{\infty} \Gamma_i$$

Then, Γ^* is a maximal non-trivial set containing Γ. Each finite subset of Γ^* must be contained in some Γ_i for some i, and thus must be non-trivia (since Γ_i is non-trivial).

For suppose that $A \in \mathbf{F}$ and $A \notin \Gamma^*$. As A is a formulas of $\mathbf{F}$, it must appear in our enumeration, say as A_k. If $\Gamma \cup \{A_k\}$ were non-trivial, then our construction would guarantee that $A_k \in \Gamma_{k+1}$, and hence $A_k \in \Gamma^*$. Because $A_k \notin \Gamma^*$, it follows that $\Gamma_k \cup \{A\}$ is also trivial. Hence $\Gamma^* \cup \{A\}$ is also trivial. It follows that Γ^* is a maximal non-trivial set.

As $\Gamma \subseteq \Gamma_i$ and $i \in \omega$, we have that $\Gamma \subseteq \Gamma^* = \bigcup_{i=0}^{\infty} \Gamma_i$. On the other hand, suppose that Γ^* is trivial. Thus, $\Gamma^* = \mathbf{F}$. Thus, $p_\lambda, \neg_* p_\lambda \in \Gamma^*$. As τ is finite, we have that any application of *modus ponens* has only a finite number of premises. Thus, there are $n, m < \omega$ such that $p_\lambda \in \Gamma_n$ and $\neg_* p_\lambda \in \Gamma_m$. Therefore, $p_\lambda, \neg_* p_\lambda \in \Gamma_{n_0}$, where $n_0 = max(n, m)$. Thus, Γ_{n_0} is trivial, but it is a contradiction.

Theorem 2.15 *Let Γ be a maximal non-trivial set of formulas. Then, we have the following:*

1. *if A is an axiom of $P\tau$, then $A \in \Gamma$*
2. *$A, B \in \Gamma$ iff $A \wedge B \in \Gamma$*
3. *$A \vee B \in \Gamma$ iff $A \in \Gamma$ or $B \in \Gamma$*
4. *if $p_\lambda, p_\mu \in \Gamma$, then $p_\theta \in \Gamma$, where $\theta = max(\lambda, \mu)$*
5. *$\neg^k p_\mu \in \Gamma$ iff $\neg^{k-1} p_{\sim\mu} \in \Gamma$, where $k \geq 1$*
6. *if $A, A \rightarrow B \in \Gamma$, then $B \in \Gamma$*
7. *$A \rightarrow B \in \Gamma$ iff $A \notin \Gamma$ or $B \in \Gamma$*

Proof We only prove (4). (5) is clear by definition. The remaining cases are proved as in the classical cases. The proof of (4) is as follows. From p_λ, p_μ, by (2), we have that (a).

(a) $p_\lambda \wedge p_\mu \in \Gamma$

From the axiom (τ_4), (b) holds.

(b) $p_\lambda \wedge p_\mu \rightarrow p_\theta$

where $\theta = max(\lambda, \mu)$. By (1), from (b), (c) follows.

(c) $p_\lambda \wedge p_\mu \rightarrow p_\theta \in \Gamma$

By (7) from (a) and (c), we can derive (d).

(d) $p_\theta \in \Gamma$

Theorem 2.16 *Let Γ be a maximal non-trivial set of formulas. Then, the characteristic function χ of Γ, that is, $\chi_\Gamma \to \mathbf{2}$ is the valuation function of some interpretation $I : \mathbf{P} \to |\tau|$.*

Proof Let us define the function $I : \mathbf{P} \to |\tau|$ putting $I(p) = \bigvee\{\mu \in |\tau| \mid p_\mu \in \Gamma\}$. Such a function is well defined, so $p_\perp \in \Gamma$.

Let $v_I : \mathbf{F} \to \mathbf{2}$ be the valuation associated to I. Hereafter, we omit the subscript. We need to show $v = \chi_\Gamma$. To show this, let $p_\mu \in \Gamma$. Thus, $\chi_\Gamma(p_\mu) = 1$. On the other hand, it is clear that $I(p) \geq \mu$. So, $v(p_\mu) = 1$. If $p_\mu \notin \Gamma$, $\chi_\Gamma(p_\mu) = 0$. Also, $I(p) \ngeq \mu$, because if so, that is $I(p) \geq \mu$, we have $p_{I(p)} \in \Gamma$, which is a contradiction. Therefore, $I(p) \ngeq \mu$, and thus $v_{p_\mu} = 0$.

By Theorem 2.15(5), $\neg^k p_\mu \in \Gamma$ iff $\neg^{k-1} p_{\sim\mu} \in \Gamma$, where $k \geq 1$. Thus, $\chi_\Gamma(\neg^k p_\mu) = \chi_\Gamma(\neg^{k-1} p_{\sim\mu})$, where $k \geq 1$. We show that $v(\neg^k p_\mu) = \chi_\Gamma(\neg^k p_\mu)$. We proceed by induction on k. If $k = 0$, it is just the previous case. Suppose that it holds for $k - 1$ ($k \geq 1$). Then, we have that $\chi_\Gamma(\neg^k p_\mu) = \chi_\Gamma(\neg^{k-1} p_{\sim\mu}) = v(\neg^{k-1} p_{\sim\mu}) = v(\neg^k p_\mu)$.

Now, let A be any formula. We proceed by induction on the number of occurrences of connectives in A. Thus, suppose that:
(1) A is of the form $\neg B$: Due to the previous discussion, we can suppose that B is a complex formula. So, $\chi_\Gamma(B) = v(B)$. If $A \in \Gamma$, then $B \notin \Gamma$, and $\chi_\Gamma(A) = 0$ and $\chi_\Gamma(B) = 1$. But, $v(A) = 1 - v(B)$, Therefore, $v(A) = 0$.
(2) A is of the form $B \wedge C$: $A \in \Gamma$ iff $B, C \in \Gamma$. By induction hypothesis, $\chi_\Gamma(B) = v(B)$ and $\chi_\Gamma(C) = v(C)$. Thus, $\chi_\Gamma = v(A)$. The other cases can also easily proved.

Here is the completeness theorem for $P\tau$.

Theorem 2.17 (Completeness) *Let Γ be a set of formulas and A be any formula. If τ is finite, then $\mathscr{A}\tau$ is a complete axiomatization for $P\tau$, i.e., if $\Gamma \models A$ then $\Gamma \vdash A$.*

Proof It can be proved by contraposition. Suppose that $\Gamma \nvdash A$. Thus, $\Gamma_0 = \Gamma \cup \{\neg_* A\}$ is non-trivial. By Theorem 2.14, Γ_0 is contained in a maximal non-trivial set Γ. Let $v : \mathbf{F} \to \mathbf{2}$ be the valuation obtained from Γ. We have that $v(A) = 1 - v(\neg_* A) = 0$. Thus, $\Gamma \not\models A$.

The decidability theorem also holds for finite lattice.

Theorem 2.18 (Decidability) *If τ is finite, then $P\tau$ is decidable.*

Proof Let A be a formula. We denote by $sf(A)$ the set of all subformulas of A and by $at(A)$ the set of all atomic subformulas composing A. We write $\sharp A$ for the cardinality of the set A. So, by using the valuation defined above, we can check in $\sharp sf(A) - \sharp at(A)$ steps as in the classical case up to analyze $\sharp at(A)$ atomic formulas. The validity of each atomic formula is checked in $\sharp\tau$ times. So, $at(A)$ can be checked at most $k\sharp\tau\sharp at(A)$ times, where k is a constant. Thus, it is possible to check whether A is valid or not in a finite number of steps. This means that $P\tau$ is decidable.

The completeness does not in general hold for an infinite lattice. But, it holds for a special case.

Definition 2.14 (*Finite annotation property*) Suppose that Γ be a set of formulas such that the set of annotated constants occurring in Γ is included in a finite substructure of τ (Γ itself may be infinite). In this case, Γ is said to have the *finite annotation property*.

Note that if τ' is a substructure of τ then τ' is closed under the operations $\sim$, $\vee$ and $\wedge$. One can easily prove the following from Theorem 2.17.

Theorem 2.19 (Finitary Completeness) *Suppose that Γ has the finite annotation property. If A is any formula such that $\Gamma \vdash A$, then there is a finite proof of A from Γ.*

Theorem 2.19 tells us that even if the set of the underlying truth-values of $P\tau$ is infinite (countably or uncountably), as long as theories have the finite annotation property, the completeness result applies to them, i.e., $\mathscr{A}\tau$ is complete with respect to such theories.

In general, when we consider theories that do not possess the finite annotation property, it may be necessary to guarantee completeness by adding a new infinitary inference rule (ω-rule), similar in spirit to the rule used by da Costa [60] in order to cope with certain models in a particular family of infinitary language. Observe that for such cases a desired axiomatization of $P\tau$ is not finitary.

From the classical result of compactness, we can state a version of the compactness theorem.

Theorem 2.20 (Weak Compactness) *Suppose that Γ has the finite annotation property. If A is any formula such that $\Gamma \vdash A$, then there is a finite subset Γ' of Γ such that $\Gamma' \vdash A$.*

Annotated logics $P\tau$ provide a general framework, and can be used to reasoning about many different logics. Below we present some examples.

The set of truth-values $FOUR = \{t, f, \bot, \top\}$, with $\neg$ defined as: $\neg t = f, \neg f = t, \neg\bot = \bot, \neg\top = \top$. Four-valued logic based on $FOUR$ was originally due to Belnap [49, 50] to model internal states in a computer. Subrahmanian [149] formalized an annotated logic with $FOUR$ as a foundation for paraconsistent logic programming; also see Blair and Subrahmanian [53]. In Chap. 6, we will give a detailed exposition of paraconsistent logic programming.

Their annotated logic may be used for reasoning about inconsistent knowledge bases. For example, we may allow logic programs to be finite collections of formulas of the form:

$$(A : \mu_0) \leftrightarrow (B_1 : \mu_1) \& \ldots \& (B_n : \mu_n)$$

where A and B_i ($1 \leq i \leq n$) are atoms and μ_j ($0 \leq j \leq n$) are truth-values in $FOUR$.

Intuitively, such programs may contain “intuitive” inconsistencies–for example, the pair

$$((p : f), (p : t))$$

is inconsistent. If we append this program to a consistent program P, then the resulting union of these two programs may be inconsistent, even though the predicate symbols p occurs nowhere in program P.

Such inconsistencies can easily occur in knowledge based systems, and should not be allowed to trivialize the meaning of a program. However, knowledge based systems based on classical logic cannot handle the situation since the program is trivial. In Blair and Subrahmanian [53], it is shown how the four-valued annotated logic may be used to describe this situation. Later, Blair and Subrahmanian's annotated logic was extended as *generalized annotated logics* by Kifer and Subrahmanian [100].

There are also other examples which can be dealt with by annotated logics. The set of truth-values *FOUR* with negation defined as boolean complementation forms an annotated logic.

The unit interval [0, 1] of truth-values with $\neg x = 1 - x$ is considered as the base of annotated logic for qualitative or fuzzy reasoning. In this sense, probabilistic and fuzzy logics could be generalized as annotated logics.

The interval $[0, 1] \times [0, 1]$ of truth-values can be used for annotated logics for evidential reasoning. Here, the assignment of the truth-value (μ_1, μ_2) to proposition p may be thought of as saying that the degree of belief in p is μ_1, while the degree of disbelief is μ_2. Negation can be defined as $\neg(\mu_1, \mu_2) = (\mu_2, \mu_1)$.

Note that the assignment of $[\mu_1, \mu_2]$ to a proposition p by an interpretation I does not necessarily satisfy the condition $\mu_1 + \mu_2 \leq 1$. This contrasts with probabilistic reasoning. Knowledge about a particular domain may be gathered from different experts (in that domain), and these experts may hold different views.

Some of these views may lead to a “strong” belief in a proposition; likewise, other experts may have a “strong” disbelief in the same proposition. In such a situation, it seems appropriate to report the existence of conflicting opinions, rather than use ad-hoc means to resolve this conflict.

The above examples can be described by annotated logics $P\tau$ or its suitable extensions. These issues will be taken up in Chap. 5.

Chapter 3
Predicate Annotated Logics $Q\tau$

Abstract This chapter studies the predicate annotated logics $Q\tau$, which can be seen as a predicate extension of $P\tau$. Their axiomatization and semantics are considered. We also prove completeness and other metatheorems.

3.1 Language

As described in Chap. 2, da Costa et al. [66] investigated propositional annotated logics $P\tau$, and suggested their predicate extension $Q\tau$ (also denoted $Q\mathscr{T}$). Details of $Q\tau$ can be found in da Costa et al. [62]; also see Abe [1].

Predicate annotated logics $Q\tau$ can be formalized as a two-sorted first-order logic. We repeat some definitions below. $\tau = \langle | \tau |, \leq, \sim \rangle$ is some arbitrary, but fixed complete lattice, with the ordering $\leq$ and the operator $\sim: | \tau | \rightarrow | \tau |$. The bottom element of this lattice is denoted by $\perp$, and top element is denoted by $\top$.

The language L^τ of $Q\tau$ is a first-order language without equality. Abe [1] introduced equality into $Q\tau$.

Definition 3.1 (*Symbols*) Primitive symbols are the following:

1. Logical connectives: $\wedge$ (conjunction), $\vee$ (disjunction), $\rightarrow$ (implication), and $\neg$ (negation)
2. Individual variables: a denumerably infinite set of variable symbols
3. Individual constants: an arbitrary family of constant symbols
4. Quantifiers: $\forall$ (for all) and $\exists$ (exists)
5. Function symbols: for each natural number $n > 0$, a collection of function symbols of arity n
6. Annotated predicate symbols: for any natural number $n \geq 0$, and any $\lambda \in \tau$, a family of annotated predicate symbols P_τ
7. Parentheses: (and)

Here, $\forall$ is called the *universal quantifier* and $\exists$ the *existential quantifier*. We define the notion of *term* as usual. Given an annotated predicate symbol P_λ of arity n and n terms $t_1, \ldots, t_n$, an *annotated atom* is an expression of the form $P_\lambda(t_1, \ldots, t_n)$.

J.M. Abe et al., *Introduction to Annotated Logics*,
Intelligent Systems Reference Library 88, DOI 10.1007/978-3-319-17912-4_3

Definition 3.2 (*Formulas*) Formulas are defined as follows:

1. An annotated atom is a formula.
2. If F is a formula, then $\neg F$ is a formula.
3. If F and G are formulas, then $F \wedge G$, $F \vee G$, $F \rightarrow G$ are formulas.
4. If F is a formula and x is an individual variable, then $\forall x F$ and $\exists x F$ are formulas.

Definition 3.3 (*Hyper-literal and complex formulas*) Hyper-literal and complex formulas are defined as follows. A formula of the form $\neg^k p_\mu(t_1, \ldots, t_n)$ $(k \geq 0)$ is called a *hyper-literal*. A formula which is not a hyper-literal is called a *complex formula*

As in $P\tau$, we may also use the formulas of the form $A \leftrightarrow B$ and $\neg_* A$ in $Q\tau$. Here, $\leftrightarrow$ denotes the *equivalence* and $\neg_*$ *strong negation*, respectively. We can also introduce the *equality*, denoted $=$, into $Q\tau$. If t and s are terms, then $s = t$ is also a formula. $s = t$ is read "s and t are equal".

3.2 Semantics

We describe a semantics for $Q\tau$, which is a variant of the semantics for standard first-order logic.

Definition 3.4 (*Interpretation*) An *interpretation* I for the language L^τ of $Q\tau$ consists of a non-empty set, denoted by $dom(I)$, and called the *domain*, together with

1. a function η_I that maps constants of L^τ to $dom(I)$
2. a function ζ_I that assigns, to each function symbol f of arity n in L^τ, a function from $(dom(I))^n$ to $dom(I)$
3. a function χ_I that assigns, to each predicate symbol of arity n in L^τ, a function from $(dom(I))^n$ to τ.

Definition 3.5 (*Variable assignment*) Suppose I is an interpretation for L^τ. Then, a *variable assignment* v for L^τ with respect to I is a map from the set of variables symbols of L^τ to $dom(I)$.

Definition 3.6 (*Denotation*) The *denotation* $d_{I,v}(t)$ of a term t with reference to an interpretation I and variable assignment v is defined inductively as follows:

1. If t is a constant symbol, then $d_{I,v}(t) = \eta(t)$.
2. If t is a variable symbol, then $d_{I,v}(t) = v(t)$.
3. If t is a function symbol, then $d_{I,v}(t) = \zeta(f)(d_{I,v}(t_1), \ldots, d_{I,v}(t_n))$.

Definition 3.7 (*Truth relation*) Let I and v be an interpretation of L^{τ} and a variable assignment with reference to I, respectively. We also suppose that A is an ordinary atom, and that F, G and H are any formulas whatsoever. Then, the *truth relation* $I, v \models A$, saying that A is true with reference to an interpretation I and variable assignment v, is defined as follows:

1. $I, v \models P_{\mu}(t_1, \ldots, t_n)$ iff $\chi_I(P)(d_{I,v}(t_1), \ldots, d_{I,v}(t_n)) \geq \mu$
2. $I, v \models \neg^k A_{\mu}$ iff $I, v \models \neg^{k-1} A_{\sim\mu}$
3. $I, v \models F \wedge G$ iff $I, v \models F$ and $I, v \models G$
4. $I, v \models F \vee G$ iff $I, v \models F$ or $I, v \models G$
5. $I, v \models F \to G$ iff $I, v \not\models F$ or $I, v \models G$
6. $I, v \models \neg F$ iff $I, v \not\models F$, where F is not a hyper-literal
7. $I, v \models \exists x H$ iff for some variable assignment v' such that for all variables y different from x, $v(y) = v'(y)$, we have that $I, v' \models H$
8. $I, v \models \forall x H$ iff for all variable assignments v' such that for all variables y different from x, $v(y) = v'(y)$, we have that $I, v' \models H$
9. $I \models H$ iff for all variable assignments v associated with I, $I, v \models H$

The equality $s = t$ is interpreted as follows:

$I, v \models s = t$ iff $d_{I,v}(s) = d_{I,v}(t)$

Here, $=$ at the right hand side of 'iff' denotes the equality symbol in the meta-language, and it reads classically. We could also introduce annotated quality $=_{\lambda}$ as a binary annotated atom. However, we do not go into details here.

As defined in Chap. 2, we can define the notions of validity, model and semantic consequence. Let $\Gamma \cup \{H\}$ be a set of formulas. We write $\models H$, and say that H is *valid* (in $Q\tau$) if, for every interpretation I, $I \models H$. If $I \models A$ for each $A \in \Gamma$, I is a *model* of Γ. We say that H is a *semantic consequence* of Γ iff for any interpretation I such that $I \models G$ for all $G \in \Gamma$, it is the case that $I \models F$.

The following lemmas concerns the properties of $\models$, whose proofs are immediate from the corresponding proof in the previous chapter.

Lemma 3.1 *For any complex formula A and B and any formula F, the valuation v satisfies the following:*

1. $\models A \leftrightarrow B$ iff $\models A \to B$ and $\models B \to A$
2. $\not\models (A \to A) \wedge \neg(A \to A)$
3. $\models \neg_* A$ iff $\not\models A$
4. $\models \neg F \leftrightarrow \neg_* F$

Lemma 3.2 *Let $A_{\mu}(t_1, \ldots, t_n)$ be an annotated atom and $\mu, \lambda \in |\tau|$. Then, we have:*

1. $\models A_{\perp}(t_1, \ldots, t_n)$
2. $\models A_{\mu}(t1, \ldots, t_n) \rightarrow A_{\lambda}(t_1, \ldots, t_n),\ \mu \geq \lambda$
3. $\models \neg^k A_{\mu}(t_1, \ldots, t_n) \leftrightarrow \neg^{k-1} A_{\sim\mu}(t_1, \ldots, t_n),\ k \geq 0$

3.3 Axiomatization

In this section, we describe a Hilbert style axiomatization of $Q\tau$, called $\mathscr{A}$. In the formulation of the postulates of $\mathscr{A}$, the symbols A, B, C denote any formula whatsoever, F and G denote complex formulas, and P_{λ} is an annotated atom.

Postulates for $\mathscr{A}$ described in Abe [1] are as follows; also see da Costa et al. [62].

Postulates for $\mathscr{A}$

$(\rightarrow_1)$ $(A \rightarrow (B \rightarrow A)$
$(\rightarrow_2)$ $(A \rightarrow (B \rightarrow C)) \rightarrow ((A \rightarrow B) \rightarrow (A \rightarrow C))$
$(\rightarrow_3)$ $((A \rightarrow B) \rightarrow A) \rightarrow A$
$(\rightarrow_4)$ $A, A \rightarrow B/B$
$(\wedge_1)$ $(A \wedge B) \rightarrow A$
$(\wedge_2)$ $(A \wedge B) \rightarrow B$
$(\wedge_3)$ $A \rightarrow (B \rightarrow (A \wedge B))$
$(\vee_1)$ $A \rightarrow (A \vee B)$
$(\vee_2)$ $B \rightarrow (A \vee B)$
$(\vee_3)$ $(A \rightarrow C) \rightarrow ((B \rightarrow C) \rightarrow ((A \vee B) \rightarrow C))$
$(\neg_1)$ $(F \rightarrow G) \rightarrow ((F \rightarrow \neg G) \rightarrow \neg F)$
$(\neg_2)$ $F \rightarrow (\neg F \rightarrow A)$
$(\neg_3)$ $F \vee \neg F$
$(\exists_1)$ $A(t) \rightarrow \exists x A(x)$
$(\exists_2)$ $A(x) \rightarrow B/\exists x A(x) \rightarrow B$
$(\forall_1)$ $\forall x A(x) \rightarrow A(t)$
$(\forall_2)$ $A \rightarrow B(x)/A \rightarrow \forall x B(x)$
(τ_1) $P_{\perp}(a_1, \ldots, a_n)$
(τ_2) $\neg^k P_{\lambda}(a_1, \ldots, a_n) \leftrightarrow \neg^{k-1} P_{\sim\lambda}(a_1, \ldots, a_n)$
(τ_3) $P_{\lambda}(a_1, \ldots, a_n) \rightarrow P_{\mu}(a_1, \ldots, a_n)$, where $\lambda \geq \mu$
(τ_4) If $A \rightarrow P_{\lambda_j}(a_1, \ldots, a_n)$, then$A \rightarrow P_{\lambda}(a_1, \ldots, a_n)$ for every $j \in J$,

where $\lambda = \bigvee_{i=1}^{m} \lambda_i$

As τ is a complete lattice, the supremum in (τ_4) is well-defined. The postulates for quantifiers are subject to the usual restrictions. When τ is finite, (τ_4) can be replaced by the schema:

$$P_{\lambda_1}(a_1, \ldots, a_n) \wedge P_{\lambda_2}(a_1, \ldots, a_n) \wedge \cdots \wedge P_{\lambda_m}(a_1, \ldots, a_n) \rightarrow P_{\lambda}(a_1, \ldots, a_n),$$

where $\lambda = \bigvee_{i=1}^{m} \lambda_i$

Here, $(\rightarrow_4)$, $(\exists_4)$, $(\forall_4)$ and (τ_4) are regarded as rules of inference

Abe [1] also added the following three axioms for equality:

$(=_1)\ x = x$
$(=_2)\ x_1 = y_1 \rightarrow (\cdots \rightarrow (x_n = y_n \rightarrow f(x_1, \ldots, x_n) = f(y_1, \ldots, y_n)))$
$(=_3)\ x_1 = y_1 \rightarrow (\cdots \rightarrow (x_n = y_n \rightarrow P(x_1, \ldots, x_n) \rightarrow P(y_1, \ldots, y_n)))$

Here, f and P are function symbol and predicate symbol, respectively.

As in $\mathscr{A}\tau$, we easily define the syntactic concepts related to $\mathscr{A}$; in particular the concepts of syntactic consequence $\vdash$ is defined in the normal way. We only note that the notion of deduction (proof) is not finitary if τ is infinite.

da Costa, Abe and Subrahmanian's axiomatization of $\mathscr{A}$ adopts different naming for postulates, but it is equivalent to the above axiomatization. The deduction theorem (Theorem 2.8) also holds for $\mathscr{A}$.

Theorem 3.1 *The following dualities of quantifiers hold:*

1. $\vdash \forall x A \leftrightarrow \neg_* \exists x \neg_* A$
2. $\vdash \exists x A \leftrightarrow \neg_* \forall x \neg_* A$

3.4 Formal Results

In this section, we give some formal results of $Q\tau$. The first result is soundness of $Q\tau$.

Theorem 3.2 (Soundness) *Let $\Gamma \cup \{A\}$ be a set of formulas of $Q\tau$. Then, $\Gamma \vdash A$ (in $\mathscr{A}$) implies that $\Gamma \vdash A$, i.e., $\mathscr{A}$ is sound with respect to the semantics of $Q\tau$.*

Proof By induction on the length of deduction. (When τ is infinite, the reasoning is by transfinite induction.)

The next result is completeness of $Q\tau$ in a restricted sense.

Theorem 3.3 (Completeness) *Let $\Gamma \cup \{A\}$ be a set of formulas of $Q\tau$. Then, if τ is finite or if $\Gamma \cup \{A\}$ possesses the finite annotation property, we have that $\Gamma \models A$ entails $\Gamma \vdash A$, i.e., $\mathscr{A}$ is complete with respect to the semantics of $Q\tau$.*

Proof By a simple extension of the proof of completeness of $P\tau$, which was described in the previous chapter.

When τ is infinite, it seems that completeness can be obtained only by augmenting $\mathscr{A}$ with an extra infinitary rule.

$Q\tau$ belong to the class of non-classical logics, and they are paraconsistent and paracomplete. They have a weak negation $\neg$, but we can define the strong negation $\neg_*$, which is classical.

da Costa et al. [62] presented another axiomatization of $Q\tau$ with a different nature, which is obtained by adjoining to the classical first-order logic, a weak negation $\neg$ plus some extra convenient postulates.

Let $\mathscr{C}$ be an axiomatization of classical first-order logic (without equality), in which negation is denoted by $\sim$. The remaining primitives defined symbols of $\mathscr{C}$ are the same as the corresponding one of $Q\tau$. We also suppose that the atomic formulas of the language of $\mathscr{C}$ are annotated atoms, as in L^τ. Furthermore, we suppose that we have added to $\mathscr{C}$ a weak negation $\neg$.

We denote by $\mathscr{A}'$ the axiomatic system obtained from $\mathscr{C}$ by adding the axioms $(\neg_1)$, $(\neg_2)$, $(\neg_3)$, (τ_1), (τ_2), (τ_3), (τ_4), and the rule:

If F and G are formulas such that G is obtained from F by the replacement of a sub-formula of the form $\neg_* A$ by $A \to (A \to A) \wedge \neg(A \to A)$ or by the replacement of a sub-formula of the latter from by one of the first form, then infer $F \leftrightarrow G$.

Theorem 3.4 *$\mathscr{A}$ and $\mathscr{A}'$ are equivalent; both characterize $Q\tau$.*

Proof Any postulate of $\mathscr{A}$ is a postulate of $\mathscr{A}'$ or is provable in $\mathscr{A}'$, and the definition $\neg_* A =_{def} A \to (A \to A) \wedge \neg(A \to A)$ corresponds to a rule in $\mathscr{A}'$. Conversely, any postulate of $\mathscr{A}'$ is a postulate or definition of $\mathscr{A}$ or is provable in $\mathscr{A}$. Therefore, $\mathscr{A}$ and $\mathscr{A}'$ are equivalent.

Theorem 3.4 reveals that annotated logic $Q\tau$ can be interpreted as an extension of classical first-order logic C. This fact seems interesting theoretically as well as practically.

Annotated logics can be used for various mathematical subjects. For example, it is possible to work out a set theory based on $Q\tau$. We will explore annotated set theory in Chap. 4. For this purpose, we need the notion of normal structure, and need to define a fragment of $Q\tau$.

Definition 3.8 (*Normal structure*) Let X be a non-empty set. A *normal structure* based on X is a function $f : X \times X \to \tau$.

We denote by $Q\tau^2$ the logic $Q\tau$ obtained by suppressing all function symbols and all predicate symbols, with the exception of one predicate symbol of arity 2 (a binary predicate symbol) which we represent by $\in$. $A\tau^2$ is then a dyadic predicate calculus whose atoms are annotated by τ. An annotated atom of $Q\tau^2$ has the form $\in_\lambda (a, b)$, where a and b are terms and $\lambda \in \tau$. This atom will be written $a \in_\lambda b$.

Intuitively, $\in$ is the membership predicate symbol. The subscript λ denotes a "degree" of membership. A normal structure is basically just a first-order interpretation as defined earlier with the following differences. First, $Q\tau^2$ contains only one predicate symbol $\in$ associated with different members of τ. Second, the normal structures are the interpretations of $\in$.

Theorem 3.5 *$Q\tau^2$ is sound with respect to the semantics of normal structures. If τ is finite or we consider only sets of formulas sharing the finite annotation property, then $Q\tau^2$ is also complete.*

Proof Consequence of Theorems 3.2 and 3.3.

We have completed the basics of annotated logics $P\tau$ and $Q\tau$. Other formal issues will be discussed in detail in the next chapter.

Chapter 4
Formal Issues

Abstract This chapter discusses formal issues of annotated logics. We describe an algebraic semantics for $P\tau$ based on Curry algebras. We also discuss annotated set theory, annotated model theory, proof methods, and annotated modal logics.

4.1 Algebraic Semantics

We gave a standard model-theoretic semantics for annotated logic. For $P\tau$, we can also provide an *algebraic semantics*. Algebraic semantics is mathematically more elegant than model-theoretic semantics. However, algebraic semantics for paraconsistent logics challenges standard formulation, since known techniques cannot be properly used.

In this section, we describe an algebraic semantics for $P\tau$ by introducing its algebraic version following Abe [4]. For this purpose, we rely on the so-called *Curry algebra* due to Barros et al. [42]. We introduce Curry algebra $P\tau$ that algebraizes propositional annotated logics $P\tau$. Systems induced by Curry algebra are called *Curry systems*, which constitute a very promising field of research but few people have paid attention to the subject.

In order to obtain algebraic versions of the majority of logical systems the procedure is the following: we define an appropriate equivalence relation in the set of formulas (e.g. identifying equivalent formulas in classical propositional logic), in such a way that the primitive connectives are compatible with the equivalence relation, i.e., a congruence.

The resulting quotient system is the algebraic structure linked with the corresponding logical system. By this process, Boolean algebra constitutes the algebraic version of classical propositional logic, Heyting algebra constitutes the algebraic version of intuitionistic propositional logic, and so on. Thus, the procedure is to formulate an algebraic semantics.

However, in some non-classical logics, it is not always clear what "appropriate" equivalence relation here can be; the non-existence of any significant equivalence relation among formulas of the calculus can also take place. This occurs, for instance, with some paraconsistent systems; see Mortensen [122]. Indeed, as pointed

J.M. Abe et al., *Introduction to Annotated Logics*,
Intelligent Systems Reference Library 88, DOI 10.1007/978-3-319-17912-4_4

out by Eytan [79], even for classical logic, it may not always be convenient to apply these ideas.

Now, we give some basic definitions related to Curry algebras. In $P\tau$, we define $A \leq B$ by setting $\vdash A \rightarrow B$, and $A \equiv B$ by setting $A \leq B$ and $B \leq A$. Here, $\leq$ is a quasi-order and $\equiv$ is an equivalence relation, respectively. Let R be a set whose elements are denoted by x, y, z, x', y'.

Definition 4.1 (*Curry pre-ordered system*) A system $(R, \equiv, \leq)$ is called a *Curry pre-ordered system*, if

1. $\equiv$ is an equivalence relation on R
2. $x \leq x$
3. $x \leq y$ and $y \leq z$ imply $x \leq z$
4. $x \leq y, x' \equiv$ and $y' \equiv y$ imply $x' \leq y'$.

Definition 4.2 (*Pre-lattice*) A system $(R, \equiv, \leq)$ is called a *pre-lattice*, if $(R, \equiv, \leq)$ is a Curry pre-ordered system and

1. $\inf\{x, y\} \neq \emptyset$
2. $\sup\{x, y\} \neq \emptyset$.

We denote by $x \wedge y$ one element of the set of $\inf\{x, y\}$ and by $x \vee y$ one element of the set of $\sup\{x, y\}$.

Definition 4.3 (*Implicative pre-lattice*) A system $(R, \equiv, \leq)$ is called a *implicative pre-lattice*, if

1. $(R, \equiv, \leq)$ is a pre-lattice
2. $x \wedge (x \rightarrow y) \leq y$
3. $x \wedge y \leq z$ iff $x \leq y \rightarrow z$.

Definition 4.4 An *implicative pre-lattice* $(R, \equiv, \leq)$ is called *classic* if $(x \rightarrow y) \rightarrow x \leq y$ (Peirce's law).

As is obvious from the above definitions, a classic implicative pre-lattice is a pre-algebraic structure which can characterize positive classical propositional logic, i.e., classical propositional logic without negation. As is well known, Peirce's law corresponds to the law of excluded middle.

We are now ready to define a Curry algebra $P\tau$. Let S be a non-empty set and $\tau = (|\tau|, \leq)$ be a finite lattice with the operation $\sim: |\tau| \rightarrow |\tau|$. We denote by S^* the set of all pairs (p, λ), where $p \in S$ and $\lambda \in |\tau|$.

We now consider the set $S^* \cup \{\neg, \wedge, \vee, \rightarrow\}$. Let S^{**} be the smallest algebraic structure freely generated by the set $S^* \cup \{\neg, \wedge, \vee, \rightarrow\}$ by the usual algebraic method. Elements of S^{**} are classified in two categories: *hyper-literal elements* are of the form $\neg^k(p, \lambda)$ and *complex elements* are the remaining elements of S^{**}.

Now, we introduce the concept of a Curry algebra $P\tau$.

Definition 4.5 (*Curry algebra $P\tau$*) A *Curry algebra $P\tau$* (abbreviated by $P\tau$-algebra) is a structure $R\tau = (R, (|\tau|, \leq, \sim), \equiv, \rightarrow, \neg)$ and, for $p \in R, a \in R^*, x, y \in R^{**}$:

1. R^{**} is a classical implicative lattice with a greatest element 1
2. $\neg$ is a unary operator $\neg : R^{**} \rightarrow R^{**}$
3. $x \rightarrow y \leq (x \rightarrow \neg y) \rightarrow \neg x$
4. $x \leq \neg x \rightarrow a$
5. $p_{\perp} \equiv 1$
6. $x \vee \neg x \equiv 1$
7. $\neg^k(p, \lambda) \equiv \neg^{k-1}(p, \sim \lambda), k \geq 1$
8. If $\mu \leq \lambda$ then $(p, \mu) \leq (p, \lambda)$
9. $(p, \lambda_1) \wedge (p, \lambda_2) \wedge \cdots \wedge (p, \lambda_n) \leq (p, \lambda)$, where $\lambda = \bigvee_{i=1}^{n} \lambda_i$

One can easily see that a $P\tau$-algebra is distributive and has a greatest element as well as a first element.

Definition 4.6 Let x be an element of a $P\tau$-algebra. We put:

$$\neg_* x = x \rightarrow ((x \rightarrow x) \wedge \neg(x \rightarrow x))$$

In a $P\tau$-algebra, $\neg_* x$ is a Boolean complement of x, so both $x \vee \neg_* x \equiv 1$ and $x \wedge \neg_* x \equiv 0$ hold.

Theorem 4.1 *In a $P\tau$-algebra, the structure composed by the underlying set and by operations $\wedge$, $\vee$ and $\neg_*$ is a pre-Boolean algebra. If we pass to the quotient through the basic relation $\equiv$, we obtain a Boolean algebra in the usual sense.*

Proof A *pre-Boolean algebra* is a partial preorder $(R, \leq)$ such that the quotient by the relation $\equiv$. Thus, by definition of $P\tau$-algebra, the mentioned structure is a pre-Boolean algebra. In addition, replacing the class of equivalent formulas by a formula can produce a usual Boolean algebra in which the meet $\wedge$ is conjunction, the join $\vee$ is disjunction, and the complement is negation.

Definition 4.7 Let $(R, (|\tau|, \leq, \sim), \equiv, \leq, \rightarrow, \neg)$ be a $P\tau$-algebra, and $(R, (|\tau|, \leq, \sim), \equiv, \leq, \rightarrow, \neg_*)$ the Boolean algebra that is isomorphic to the quotient algebra of $(R, (|\tau|, \leq, \sim), \equiv, \leq, \rightarrow, \neg_*)$ by $\equiv$ is called the Boolean algebra *associated with* the $P\tau$-algebra.

Hence, we can establish the following first representation theorem for $P\tau$-algebra.

Theorem 4.2 *Any $P\tau$-algebra is associated with a field of sets. Moreover, any $P\tau$-algebra is associated with the field of sets simultaneously open and closed of a totally disconnected compact Hausdorff space.*

Proof It follows from the aforementioned definition (Definition 4.7) and the classical representation theorem for Boolean algebra by Stone which states that every Boolean algebra is isomorphic to the Boolean algebra of all open and closed sets of a (totally disconnected compact Hausdorff) topological space.

This is not the only way of extracting Boolean algebra out of $P\tau$-algebra. There is another natural Boolean algebra associated with a $P\tau$-algebra.

Definition 4.8 Let $(R, (|\tau|, \leq, \sim), \equiv, \leq, \rightarrow, \neg)$ be a $P\tau$-algebra. By RC we indicate the set of all complex elements of $(R, (|\tau|, \leq, \sim), \equiv, \leq, \rightarrow, \neg)$.

Then, the structure $(RC, (|\tau|, \leq, \sim), \equiv, \leq, \rightarrow, \neg)$ constitutes a pre-Boolean algebra which we call Boolean algebra *c-associated with* the $P\tau$-algebra $(R, (|\tau|, \leq, \sim), \equiv, \leq, \rightarrow, \neg)$. Thus, we obtain a second representation theorem for $P\tau$-algebra.

Theorem 4.3 *Any $P\tau$-algebra is c-associated with a field of sets. Moreover, any $P\tau$-algebra is c-associated with the field of sets simultaneously open and closed of a totally disconnected compact Hausdorff space.*

Proof It follows from the definition of Boolean algebra c-associated with the $P\tau$-algebra and Stone's representation theorem for Boolean algebra.

Theorems 4.7 and 4.8 show us that $P\tau$-algebra constitute interesting generalizations of the concept of Boolean algebra. There are some open questions related to these results. How many non-isomorphic Boolean algebra associated with a $P\tau$-algebra is there? How many non-isomorphic Boolean algebra c-associated with a $P\tau$-algebra is there? The answers to these questions can establish connections of associated and c-associated algebra.

Next, we show soundness and completeness of $P\tau$-algebras using the notion of filter and ideal of a $P\tau$-algebra.

Definition 4.9 (*Filter*) Let $(R, (|\tau|, \leq, \sim), \equiv, \leq, \rightarrow, \neg)$ be a $P\tau$-algebra. A subset F of R is called a *filter* if:

1. $x, y \in F$ imply $x \wedge y \in F$
2. $x \in F$ and $y \in R$ imply $x \vee y \in F$
3. $x \in F, y \in R$, and $x \equiv y$ imply $y \in F$.

Definition 4.10 (*Ideal*) Let $(R, (|\tau|, \leq, \sim), \equiv, \leq, \rightarrow, \neg)$ be a $P\tau$-algebra. A subset I of R is called an *ideal* if:

1. $x, y \in I$ imply $x \vee y \in I$
2. $x \in I$ and $y \in R$ imply $x \wedge y \in F$
3. $x \in I$, $y \in R$, and $x \equiv y$ imply $y \in F$.

Then, we have the following lemma whose proof is trivial.

Lemma 4.1 *Let $(R, (|\tau|, \leq, \sim), \equiv, \leq, \rightarrow, \neg)$ be a $P\tau$-algebra. A subset F of R is a filter iff:*

1. $x, y \in F$ *imply* $x \wedge y \in F$
2. $x \in F, y \in R$, *and* $x \leq y$ *imply* $y \in F$
3. $x \in F, y \in R$, *and* $x \equiv y$ *imply* $y \in F$.

A subset I of R is an ideal iff:

1. $x, y \in I$ *imply* $x \vee y \in I$
2. $x \in I, y \in R$, *and* $x \leq y$ *imply* $y \in I$
3. $x \in I, y \in R$, *and* $x \equiv y$ *imply* $y \in I$.

Filters are partially ordered by inclusion. Filters that are maximal with respect to this ordering are called *ultrafilters*. By the Ultrafilter Theorem, every filter in $P\tau$-algebra can be extended to an ultrafilter.

Theorem 4.4 *Let F be an ultrafilter in a $P\tau$-algebra. Then, we have:*

1. $x \wedge y \in F$ *iff* $x \in F$ *and* $y \in F$
2. $x \vee y \in F$ *iff* $x \in F$ *or* $y \in F$
3. $x \rightarrow y \in F$ *iff* $x \notin F$ *or* $y \in F$
4. *If* $p_{\lambda_1}, p_{\lambda_2} \in F$, *then* $p_\lambda \in F$, *where* $\lambda = \lambda_1 \vee \lambda_2$
5. $\neg^k p_\lambda \in F$ *iff* $\neg^{k-1} p_{\sim\lambda} \in F$
6. *If* $x, x \rightarrow y \in F$, *then* $y \in F$

Proof We prove only (4). In fact, if $p_{\lambda_1}, p_{\lambda_2} \in F$, then by Definition 4.5(9) and Lemma 4.1, it follows that $p_\lambda \in F$, where $\lambda = \lambda_1 \vee \lambda_2$.

Definition 4.11 If $R\tau_1 = (R_1, (|\tau_1|, \leq_1, \sim_1), \equiv_1, \leq_1, \rightarrow_1, \neg_1)$ and $R\tau_2 = (R_2, (|\tau_1|, \leq_2, \sim_2), \equiv_2, \leq_2, \rightarrow_2, \neg_2)$ are two $P\tau$-algebras, then a homomorphism of $R\tau_1$ into $R\tau_2$ is a map f of R_1 into R_2 which preserves the algebraic operations, i.e., such that for $x, y \in R_1$:

1. $x \leq_1 y$ iff $f(x) \leq_2 f(y)$
2. $f(x \rightarrow_1 y) \equiv_2 f(x) \rightarrow_2 f(y)$
3. $f(\neg_1 x) \equiv_2 \neg_2 f(x)$

4. If $x \equiv_1 y$, then $f(x) \equiv_2 f(y)$
5. f is also extended to a homomorphism of $(|\tau_1|, \leq_1, \sim_1)$ into $(|\tau_2|, \leq_2, \sim_2)$ in an obvious way (i.e., for instance, $f(\sim_1 \lambda) = \sim_2 f(\lambda)$).

Then, as in the classical case, we can present the following theorem:

Theorem 4.5 *Let $R\tau_1$ and $R\tau_2$ be two $P\tau$-algebras and f a homomorphism from $R\tau_1$ into $R\tau_2$. Then, the set $\{x \in R_1 \mid f(x) \equiv_2 1_2\}$ (the shell of f) is a filter and the set $\{x \in R_2 \mid f(x) \equiv_2 0_2\}$ (the kernel of f) is an ideal.*

Theorem 4.6 *If the shell of a homomorphism f of $P\tau$-algebra is an ultrafilter, then*

1. $f(x) \equiv 1$ *and* $f(y) \equiv 1$ *iff* $f(x \wedge y) = 1$
2. $f(x) \equiv 1$ *or* $f(y) \equiv 1$ *iff* $f(x \vee y) = 1$
3. $f(x) \equiv 0$ *or* $f(y) \equiv 1$ *iff* $f(x \rightarrow y) = 1$

Proof It is a simple consequence of Theorem 4.4.

Definition 4.12 Let **F** be the set of all formulas of the propositional annotated logic $P\tau$ and f a homomorphism from **F** (considered as a $P\tau$-algebra) into an arbitrary $P\tau$-algebra. We write $f \models \Gamma$, where Γ is a subset of **F**, if for each $A \in \Gamma$, $f(A) \equiv 1$. $\Gamma \models A$ means that for all homeomorphisms f from **F** into an arbitrary $P\tau$-algebra, if $f \models \Gamma$, then $f(A) \equiv 1$.

Based on the above results, we can establish algebraic soundness and completeness of the propositional annotated logic $P\tau$.

Theorem 4.7 (Soundness) *If A is a provable formula of $P\tau$, i.e., $\vdash A$, then $f(A) \equiv 1$ for any homomorphism f from* **F** *(considered as a $P\tau$-algebra) into an arbitrary $P\tau$-algebra.*

Proof By induction on the length of proofs.

To prove completeness, we need the following theorem:

Theorem 4.8 *Let U be an ultrafilter in* **F**. *Then, there is a homomorphism f from* **F** *into* **2** $= \{0, 1\}$ *such that the shell of f is U.*

Proof Let **P** be the set of propositional variables of $P\tau$. Let us define a function $I : \mathbf{P} \rightarrow |\tau|$. If p is a propositional variable, then we put $I(p) = \bigvee\{\lambda \in |\tau| \mid p_\lambda \in U\}$. Such a function is well defined, so $p_\perp \in U$. Let $v_I : \mathbf{F} \rightarrow \mathbf{2}$ be the valuation function associated to this interpretation. We assert that $v_I = \chi_U$, which is the characterization function of U.

If $p_\lambda \in U$, then $\chi_U(p_\lambda) = 1$. On the other hand, it is clear that $I(p) \geq \lambda$. Thus, $v_I(p_\lambda) = 1$. If $p_\lambda \notin U$, then we have $\chi_U(p_\lambda) = 0$. If $I(p) \geq \lambda$, then we have $p_{I(p)} \in U$ and as $p_{I(p)} \rightarrow p_\lambda$ is an axiom, it follows that $p_\lambda \in U$, which is contradictory. So, it is not the case that $I(p) \geq \lambda$, and so $v_I(p_\lambda) = 0$.

By Theorem 4.4(5), $\neg^k p_\lambda \in U$ iff $\neg^{k-1} p_{\sim\lambda}$. Let us show that $\chi_U(\neg^k p_\lambda) = v_I(\neg^k p_\lambda)$. We proceed by induction on k. If $k = 0$, then it is just the above case. Suppose that the property hold for $k - 1$. Then, $\chi_U(\neg^k p_\lambda) = \chi_U(\neg^{k-1} p_{\sim\lambda}) = v_I(\neg^{k-1} p_{\sim\lambda}) = v_I(\neg^k p_\lambda)$.

Consider arbitrary formula A. If A is an atomic formula, then the property is valid. For other cases, we only consider the cases in which $A = \neg B$ and $A = B \wedge C$. Other cases run in a similar way. For $A = \neg B$, suppose B is a complex formula. So, $\chi_U(B) = v_I(B)$. If $A \in U$, then $B \notin U$, $\chi_U(A) = 0$ and $\chi_U(B) = 1$. But $v_I(A) = 1 - v_I(B)$. So, $v_I(A) = 0$.

For $A = B \wedge C$, $A \in U$ iff $B \in U$ and $C \in U$. By inductive hypothesis, $\chi_U(B) = v_I(B)$ and $\chi_U(C) = v_I(C)$. So, $\chi_U(A) = v_I(A)$.

Theorem 4.9 (Completeness) *Let* **F** *be the set of all formulas of the propositional annotated logic* $P\tau$ *and* $A \in$ **F**. *Suppose that* $f(A) \equiv 1$ *for any homomorphism* f *from* **F** *(considered as a* $P\tau$*-algebra) into an arbitrary* $P\tau$*-algebra. Then,* A *is a provable formula of* $P\tau$*, i.e.,* $\vdash A$.

Proof If A is not provable, then it is not the case that $A \equiv 1$ and so, it is not the case that $\neg_*(A) \equiv 0$. Therefore, there is an ultrafilter U in **F** that contains $\neg_* A$. By the previous theorem, there is a homomorphism f from **F** into **2**, and thus $f(\neg_* A) \equiv 1$. It follows that $f(A) \equiv 0$, which is a contradiction. This completes the proof.

Theorem 4.9 gives an alternative completeness result of propositional annotated logics $P\tau$ using Curry algebras $P\tau$. Curry algebras can also be applied to the completeness proof of other paraconsistent logics.

An algebraic semantics for predicate annotated logics $Q\tau$ is also interesting. Curry algebra could be expanded for $Q\tau$ in some ways. One possibility is to use a *monadic Curry algebra* which is a generalization of a Curry algebra $P\tau$ with additional operators describing two quantifiers. The initial work in this direction may be found in Abe et al. [10, 14].

4.2 Annotated Set Theory

In Chap. 3, we showed that normal structures are intimately related to annotated logic, particularly when $Q\tau$ is envisaged as an extension of classical first-order logic. *Annotated set theory*, which concerns normal structures, can thus be regarded as a generalization of classical set theory.

The most convenient way to study normal structures is to start with a classical set theory, for instance, Zermelo-Fraenkel set theory ZF and to treat them inside ZF. If we proceed this way, then annotated set theory constitutes a natural and immediate extension of fuzzy set theory.

In order to develop annotated set theory, we add to ZF (see for example Cohen [58]) two individual constants τ and $\mathscr{U}$, as well as the following axioms:

(A1) τ is a complete lattice under arbitrary, but fixed ordering $\leq$ with $\perp$ as its least element and $\top$ as its greatest element.
(A2) $\tau \in \mathscr{U}$ and $\forall X (X \in \mathscr{U} \rightarrow X \subset \mathscr{U})$.

The second axiom (A2) guarantees that $\mathscr{U}$ is a *transitive set* (X is a transitive set iff $\forall Y (Y \in X \rightarrow Y \subset X)$). In most applications, it is usually enough to postulate that τ is a set in which is defined a reflexive binary relation possessing a unique least element and a unique greatest element.

Definition 4.13 e is said to be a *normal structure based on* $\mathscr{U}$ if e is a function $\mathscr{U} \times \mathscr{U}$ into τ.

Definition 4.14 Let e be a normal function. Then, instead of $e(x, u) = \lambda$, we write $x \in_\lambda y$.

Definition 4.15 If $X \in \mathscr{U}$, then

$X^{(\lambda,e)} = \{Y \mid Y \in \mathscr{U} \text{ and } Y \in_\lambda X\}$
$X^{[\lambda,e]} = \{Y \mid Y \in \mathscr{U} \text{ and } \exists \mu (\mu \in \tau \text{ and } \mu \leq \lambda \text{ and } Y \in_\mu X)\}$
$\mathscr{F}_X = \{f \mid X \rightarrow \tau \text{ and } \exists e\, (e \text{ is a normal function and } \forall \lambda \forall Y (\lambda \in \tau$ and $T \in \mathscr{U} \rightarrow (f(Y) = \lambda \leftrightarrow Y_{e_\lambda})))\}$.

Definition 4.16 If $X, Y \in \mathscr{U}$ and $\lambda \in \tau$, then

$$X =^e_\lambda Y =_{def} \forall Z \in \mathscr{U} (Z \in_\lambda X \leftrightarrow Z \in_\lambda Y).$$

Definition 4.17 A set $X \neq \emptyset$ is said to be *strongly transitive* if it is transitive and

$$\forall Y (Y \in X \rightarrow \wp(Y) \in X).$$

where $\wp(Y)$ is the powerset of Y.

Definition 4.18 X is called a *universe* iff it is strongly transitive and for every function $f : Y \rightarrow X$ such that $Y \in X$, it is the case that $\bigcup range(f) \in X$, where $range(f) = \{z \mid \exists t (t \in Y \text{ and } f(t) = z\}$.

Theorem 4.10 *If X is a universe, then X is a standard model of all axioms of ZF, with the possible exception of the axiom of infinity.*

Proof Clearly, extensionality, choice and regularity are satisfied in X, since they are satisfied in ZF. On the other hand, if the conditions of Definitions 4.17 and 4.18 are satisfied, we easily see that the axioms of pair, union, power-set and (separation, in particular) are true in X.

Theorem 4.11 *X is a complete universe iff $\omega \in X$.*

Proof By Theorem 4.10, X satisfies all axioms of ZF with the possible exception of the axiom of infinity. But since X is a complete universe, $\omega \in X$ and this last axiom is satisfied.

When we are handling the normal functions based on X, we have to use set-theoretic constructs. The more such constructs exist, the better. So in most cases, it is convenient to suppose that $\mathcal{U}$ is a universe, i.e., a model of ZF. For example, we have:

Theorem 4.12 *If $\mathcal{U}$ is a universe, $\lambda \in \tau$, and $X \in \mathcal{U}$, then $X^{[\lambda,e]} \in \mathcal{U}$ and $X^{(\lambda,e)} \in \mathcal{U}$ and $\mathcal{F}_X \in \mathcal{U}$. Furthermore, $\{Y \mid F(Y)$ and $Y \in_\lambda X\} \in \mathcal{U}$, where $F(Y)$ is any formula of ZF, and $\forall X \forall Y(X, Y \in \mathcal{U} \rightarrow \forall e \forall \lambda$ (e is a normal function and $\lambda \in \tau)$ $\rightarrow X =^e_\lambda Y) \leftrightarrow X = Y))$.*

Proof Immediate, since $\mathcal{U}$ is model of ZF possibly minus the axiom of infinity.

The introduction of a weak negation $\neg$ in the theory of annotated set offers no difficulty. For example, we may introduce it so as to apply only to hyper-literal, and put as a postulate that

$$\neg^k(X \in_\lambda Y) \leftrightarrow \neg^{k-1}(X \in_{\sim\lambda} Y)$$

We turn to some applications of annotated set theory. To begin with, we show how the notion of *fuzzy set* can be subsumed by our technical development. Fuzzy set was proposed by Zadeh [159] to formalize fuzzy concepts and later has many applications in various areas.

Suppose $\mathcal{U}$ is a set. Then, a fuzzy set is defined as follows:

Definition 4.19 (*Fuzzy set*) A *fuzzy set* of $\mathcal{U}$ is a function $u : \mathcal{U} \rightarrow [0, 1]$. $\mathcal{F}_\mathcal{U}$ will denote the set of all fuzzy sets of $\mathcal{U}$.

Several operations on fuzzy sets are defined as follows:

Definition 4.20 For all $u, v \in \mathcal{F}_\mathcal{U}$ and $x \in \mathcal{U}$, we put

$$(u \vee v)(x) = \sup\{u(x), v(x)\}$$
$$(u \wedge v)(x) = \inf\{u(x), v(x)\}$$
$$\overline{u}(x) = 1 - u(x)$$

Definition 4.21 Two fuzzy sets $u, v \in \mathcal{F}_\mathcal{U}$ are said to be *equal* iff for every $x \in \mathcal{U}$, $u(x) = v(x)$.

Definition 4.22 $\mathbf{1}_\mathcal{U}$ and $\mathbf{0}_\mathcal{U}$ are the fuzzy sets of $\mathcal{U}$ such that for all $x \in \mathcal{U}$, $\mathbf{1}_\mathcal{U} = 1$ and $\mathbf{0}_\mathcal{U} = 0$.

It is easy to prove that $\langle \mathscr{F}_{\mathscr{U}}, \wedge, \vee \rangle$ is a complete lattice having infinite distributive property. Furthermore, $\langle \mathscr{F}_{\mathscr{U}}, \wedge, \vee, ^- \rangle$ constitutes an algebra, which in general is not Boolean (for details, see Negoita and Ralescu [131]).

A fuzzy set u of $\mathscr{U}$ can be identified with a normal structure $\overline{u}$ based on the set $\mathscr{U} \cup \{0, 1\}$ such that $\tau = \{\perp, \top\}$ and

$$\overline{u} = \begin{cases} \top \text{ if } x \in \mathscr{U} \text{ and } y = u(x) \\ \perp \text{ otherwise} \end{cases}.$$

Hence, the theory of normal structures encompasses that of fuzzy sets. It is clear that if $\mathscr{U}$ is a universe, then the definition of fuzzy set in terms of normal functions can be simplified.

Similarly, the theory of *L-fuzzy sets* and *flou sets* can be simplified as particular cases of the general theory of normal functions when we extend a little, the concept of normal function. On the theories of flou sets and of L-fuzzy sets; see Negoita and Ralescu [131].

4.3 Annotated Model Theory

It is possible to analyze annotated logics by well known model-theoretic techniques. In fact, *annotated model theory* as sketched here shows that practically all classical results can be adapted to predicate annotated logics $Q\tau$. In Chap. 3, we proved completeness of $Q\tau$. Here, we establish some necessary results for the development of the model theory of $Q\tau$.

Theorem 4.13 (Equivalence theorem) *Let $B_1, \ldots, B_n$ and $B'_1, \ldots, B'_n$ be complex formulas. Let A' be the formula obtainable from A by replacing some occurrences of $B_1, \ldots, B_n$ by $B'_1, \ldots, B'_n$, respectively. Then, if $\vdash B_i \leftrightarrow B'_i$ $(i = 1, \ldots, n)$, then $\vdash A \leftrightarrow A'$.*

Proof By induction on the length of A, as in the classical corresponding theorem. Nonetheless, we need to consider the case when A is of the form $\neg C$. There are two possibilities:

1. an occurrence of a formula in A in the whole of A
2. it is entirely contained in C.

The first case is immediate. For the second case. A' is $\neg C'$, where C' is obtainable from C by replacements of the type described in the statement of the theorem. By induction hypothesis, $\vdash C \leftrightarrow C'$, as C and C' are complex formulas, it follows that $\vdash \neg C \leftrightarrow \neg_* C$, and so $A \leftrightarrow A'$ by Lemma 2.6.

Recall some standard syntactic notions. A formula A is *open* if A does not contain quantifiers. A formula A is said to be in *prenex form* if it has the form $Qx_1, \ldots, Qx_n B$, where each Qx_i is either $\forall x_i$ or $\exists x_i$, $x_1, \ldots, x_n$ are distinct and B is open. We call $Qx_1, \ldots, Qx_n$ the *prefix* and B the *matrix* of A. We allow the prefix to be empty.

By using the equivalence theorem, we can show that for some (but not for all) formulas there are equivalent formulas in prenex form.

We say that a formula A is *logically equivalent* to a formula B iff $\vdash A \leftrightarrow B$. A formula is said to be *universal-existential* iff it is logically equivalent to the formula in prenex form such that all the universal quantifiers precede all the existential quantifiers in the prefix. We abbreviate this simply by a $\forall\exists$-formula.

Let A be a formula and $x_1, \ldots, x_n$ be the variables which are free in A. The formula $\forall x_1 \ldots \forall x_n A$ is called the *closure* of A.

Theorem 4.14 (Reduction theorem) *Let Γ be a set of formulas in the theory T and A be a formula of T. Then, A is a theorem of Γ in T, abbreviate $T[\Gamma]$ iff there is a theorem of T of the form $B_1 \to (B_2 \to \cdots \to (B_n \to C)\ldots)$, where each B_i is the closure of a formula in Γ.*

Proof Immediate from the deduction theorem (Theorem 2.8).

We can generalize the reduction theorem for trivialization in two ways.

Theorem 4.15 (Reduction theorem for trivialization I) *Let Γ be a non-empty set of formulas in the theory T. Then, $T[\Gamma]$ is trivial iff there is a theorem of T that is a disjunction of strong negation of closures of distinct formulas in Γ.*

Proof Similar to the corresponding classical theorem, using strong negation instead of weak negation and trivialization instead of inconsistency.

Theorem 4.16 (Reduction theorem for trivialization II) *Let Γ be a non-empty set of formulas in the theory T. Then, $T[\Gamma]$ is trivial iff there is a theorem of T that is a disjunction of negations of closures of distinct formulas in Γ.*

Proof Consequence of Lemma 2.3(4) and Theorem 4.15.

From these two theorems, we can reach the following corollaries:

Corollary 4.1 *Let A' be the closure of any formula A. Then, A is a theorem of T iff $T[\neg_* A]$ is trivial.*

Corollary 4.2 *Let B' be the closure of a complex formula B. B is a theorem of T iff $T[\neg B']$ is trivial.*

We restate the definitions of the semantics for $Q\tau$ in the terminology of theory of models; see Shoenfield [145], although they may be immediate.

Definition 4.23 (*Structure*) A *structure* S for the language of $Q\tau$ denoted L consists of the following objects:

1. A non-empty set $|S|$, called the *universe* of S. The elements of $|S|$ are called *individuals* of S.
2. For each n-ary function symbol f of L an n-ary function $f_S : |S|^n \rightarrow |S|$. (In particular, for each constant e of L, e_S us an individual of S.)
3. For each n-ary predicate symbol p of L an n-ary function $p_s : |S| \rightarrow |\tau|$.

Let S be a structure of L and Γ be a set of formulas in L. The *diagram* of S, denoted $D_\Gamma(S)$, is the theory whose language is L_S and whose non-logical axioms are the formula A in $\Gamma(S)$ such that $S(A) = 1(true)$.

If a is a free variable, then we define the individual $S(a)$ of S. We use i and j as syntactic variables which vary over names. We define a truth-value $S(A)$ for each closed formula A in L_S.

1. If A is of the form $a = b$, then $S(A) = 1$ iff $S(a) = S(b)$.
2. If A is of the form $p_\lambda(t_1, \ldots, t_n)$, then $S(A) = 1$ iff $p_S(S(t_1), \ldots, S(t_n)) \geq \lambda$.
3. If A is of the form $B \wedge C$, $B \vee C$ or $B \rightarrow C$, then
   $S(B \wedge C) = 1$ iff $S(B) = S(C) = 1$
   $S(B \vee C) = 1$ iff $S(B) = 1$ or $S(C) = 1$
   $S(B \rightarrow C) = 1$ iff $S(B) = 0$ or $S(C) = 1$.
4. If A is of the form $\neg^k p_\lambda(t_1, \ldots, t_n)$, then $S(A) = S(\neg^{k-1} p_{\sim\lambda}(t_1, \ldots, t_n))$.
5. If A is a complex formula, then $S(\neg A) = 1 - S(A)$.
6. If A is of the form $\exists x B$, then $S(A) = 1$ iff $S(B_x[i]) = 1$ for some i in L_S.
7. If A is of the form $\forall x B$, then $S(A) = 1$ iff $S(B_x[i]) = 1$ for all i in L_S.

Here, $B_x[i]$ denotes the formula obtained from A by replacing all free occurrences of x by i.

If A is a formula of L, then S-*instance* of A is a closed formula of the form $A[i_1, \ldots, i_n]$ in $L(S)$. A formula A of L is said to be *valid* in S iff $S(A') = 1$ for every S-instance A' of A. A formula A is *logically valid* iff it is valid in every structure for L, denoted $\models A$.

Let Γ be a set of formulas and A a formula. Say that A is a *semantic consequence* of Γ iff for any structure S, $S(A) = 1$ if $S(B)$ for all $B \in \Gamma$. Other required notions are defined as follows.

Definition 4.24 (*Non-triviality*) A structure S is *non-trivial* iff there is a closed annotated atom $p_\lambda(t_1, \ldots, t_n)$ such that $S(p_\lambda(t_1, \ldots, t_n)) = 0$.

Hence, a structure S is non-trivial iff there is some closed annotated atom that is not valid in S.

Definition 4.25 (*Inconsistency*) A structure S is *inconsistent* iff there is a closed annotated atom $p_\lambda(t_1, \ldots, t_n)$ such that $S(p_\lambda(t_1, \ldots, t_n)) = S(\neg p_\lambda(t_1, \ldots, t_n)) = 1$.

Hence, a structure S is inconsistent iff there is some closed annotated atom that it and its negation are both valid in S.

Definition 4.26 (*Paraconsistency*) A structure S is *paraconsistent* iff it is both inconsistent and non-trivial.

Definition 4.27 (*Paracompleteness*) A structure S is *paracomplete* iff there is a closed annotated atom $p_\lambda(t_1, \ldots, t_n)$ such that $S(p_\lambda(t_1, \ldots, t_n)) = S(\neg p_\lambda(t_1, \ldots, t_n)) = 0$.

We can prove completeness of $Q\tau$ by means of canonical structure. As immediate consequences of the completeness theorem, we obtain the compactness theorem saying that a formula A in a theory T is valid in T if is valid in some finitely axiomatized part of T. Thus, we can state that a theory T has a model iff every finitely axiomatized part has a model.

Definition 4.28 (*Isomorphism*) Let A and B be structures for L. An *isomorphism* from A to B is a bijective function $\varphi : |A| \rightarrow |B|$ such that

1. $\varphi(f_A(a_1, \ldots, a_n)) = f_B(\varphi(a_1), \ldots, \varphi(a_n))$
2. $\varphi(p_A(a_1, \ldots, a_n)) = \lambda$ iff $p_B(\varphi(a_1), \ldots, \varphi(a_n)) = \lambda$

hold for any function symbol f, for any predicate symbol p of L, for all $a_1, \ldots, a_n \in |A|$, and for all $\lambda \in |\tau|$.

Let A and B be structures for L and $\varphi : |A| \rightarrow |B|$ be a function. If i is the name of an individual $a \in |A|$, then we use t^φ to designate the name of the individual $\varphi(a) \in |B|$. If u is an expression of $L(A)$, then u^φ is the expression obtained from u by replacing each name i by i^φ.

Theorem 4.17 *Let φ be an isomorphism between structures A and B. Then, we have:*

1. *$\varphi(A(a)) = B(a^\varphi)$ for every variable-free term a of $L(A)$.*
2. *$A(\neg^k p_\lambda(a_1, \ldots, a_n)) = B((\neg^k p_\lambda(a_1, \ldots, a_n))^\varphi)$ for every closed hyper-literal $\neg^k p_\lambda(a_1, \ldots, a_n)$ of $L(A)$.*

Proof (1) is proved by induction on the length of a. For (2), we proceed by induction on k. If $k = 0$, then we have: $A(p_\lambda(a_1, \ldots, a_n)) = 1$ iff $p_A(A(a_1, \ldots, a_n)) \geq \lambda$ iff $p_B(\varphi(A(a_1), \ldots, A(a_n)))) \geq \lambda$ iff $p_B(B(a_1{}^\varphi), \ldots, B(a_n{}^\varphi)) \geq \lambda$ iff $B(p_\lambda(a_1, \ldots, a_n)) = 1$.

Now, suppose that the theorem holds for $k > 0$. Then,

$A(\neg^{k+1} p_\lambda(a_1, \ldots, a_n))$ iff $A(\neg^k p_{\sim\lambda}(a_1, \ldots, a_n))$ by axiom (τ_2)
iff $B(\neg^k p_{\sim\lambda}(a_1, \ldots, a_n))$ by induction hypothesis
iff $B(\neg^{k+1} p_\lambda(a_1, \ldots, a_n)) = 1$ by the axiom (τ_2)

Corollary 4.3 $A(F) = B(F^\varphi)$ *for every closed formula F of $L(A)$.*

Proof By induction on the length of A. If F is a hyper-literal, then the result follows from Theorem 4.17. If F is of the form $a = b$, then it is easy to see that $A(F) = 1 = B(a^{\varphi} = b^{\varphi})$. If A is of the form $B \wedge C$, $B \vee C$, $B \rightarrow C$, $\forall x B$ or $\exists x B$, the result follows as in the classical cases.

Definition 4.29 (*Elementary equivalence*) Two structures A and B are *elementary equivalent* if the same formulas of L are valid in A and B.

The definition implies that A and B are models of the same theories.

Definition 4.30 (*Embedding*) Le A and B are structures. An *embedding* of A in B is an injective mapping $\varphi : |A| \rightarrow |B|$ such that

$$(E_1)\ f_B(\varphi(a_1), \ldots, \varphi(a_n)) = \varphi(f_A(a_1, \ldots, a_n))$$
$$(E_2)\ p_B(\varphi(a_1), \ldots, \varphi(a_n)) = \lambda \text{ iff } \varphi(f_A(a_1, \ldots, a_n))$$

hold for all function symbols f, for all predicate symbols p of L, and for all $a_1, \ldots, a_n \in |A|$, and for all $\lambda \in |\tau|$.

If $|A| \subseteq |B|$ and the identity mapping from $|A|$ to $|B|$ is an embedding of A in B, then A is called a *substructure* of B and B is called an *extension* of A. If A and B are models of some theory, we sometimes say *submodel* of substructure.

Theorem 4.18 *Let A and B be structures for L and $\varphi : |A| \rightarrow |B|$ be a mapping. Then, φ is an embedding of A in B iff $A(F) = B(F^{\varphi})$ for every variable-free formula F of L.*

Proof The proof satisfying the stated conditions is just the proof of Corollary 4.3. Now, we suppose that the conditions hold. Let $a_1, \ldots, a_n$ be individuals of A, and let $i_1, \ldots, i_n$ be the names of $a_1, \ldots, a_n$. Then, it is clear that (E_1) holds. For (E_2), we have that $A(p_{\lambda}(i_1, \ldots, i_n)) = B(p_{\lambda}(i_1{}^{\varphi}, \ldots, i_n{}^{\varphi}))$ for each $\lambda \in |\tau|$. Now, let $p_A(A(i_1), \ldots, A(i_n)) = \mu$ and $p_B(B(i_1{}^{\varphi}), \ldots, B(i_n{}^{\varphi})) = \theta$. It is clear that $A(p_{\lambda}(i_1, \ldots, i_n)) = 1$; so $B(p_{\lambda}(i_1{}^{\varphi}, \ldots, i_n{}^{\varphi}))$, that is, $p_B(B(i_1{}^{\varphi}), \ldots, B(i_n{}^{\varphi})) \geq \mu$ or $\theta \geq \mu$. In a similar way, $\mu \geq \theta$. Therefore, $\theta = \mu$. This implies (E_2). The proof of (E_1) is just like the classical one.

As a corollary to Theorem 4.18, we have:

Corollary 4.4 *Let A and B be structures for L such that $|A| \subseteq |B|$. Then, A is a substructure of A iff $A(A) = B(A)$ for every variable-free formula A of $L(A)$.*

Let Γ be a set of formulas of L and A a structure for L. $\Gamma(A)$ denotes the set of A-instances of formulas in Γ. Thus, if Γ is the set of all open formulas in L, then $\Gamma(A)$ is the set of all variable-free formulas in $L(A)$; if Γ is the set of all formulas in L, then $\Gamma(A)$ is the set of all closed formulas of $L(A)$.

Let A and B be structures for L such that $|A| \subseteq |B|$, and let Γ be a set of formulas in L. We say that A is a *Γ-substructure* of B and that B is a *Γ-extension* of A if for every formula $A \in \Gamma(A)$, $A(A) = 1$ implies $B(A) = 1$.

Theorem 4.19 *Let Γ be the set of open formulas of L. Then, the strong negation of every formula in Γ is in Γ. Moreover, the same is true for $\Gamma(A)$, where A is a structure for L.*

Proof It is enough to consult the definition of strong negation and Definition of $\Gamma(A)$.

Theorem 4.20 *Let Γ be the set of open formulas of L. Then, a Γ-substructure is simply a substructure and a Γ-extension is simply an extension.*

Proof In effect, let A be a Γ-structure of B, and A a formula in $\Gamma(A)$. If $A(A) = 0$, then $A(\neg_* A) = 1$; so $B(\neg_* A) = 1$ and therefore $B(A) = 0$.

If F is the set of all formulas in L, we say *elementary structure* for F-substructure and *elementary extension* for F-extension.

Theorem 4.21 *If $x = e$ is in Γ and A is a Γ-substructure of B, then $A(e) = B(e)$.*

Let A and B be structure for L such that $| A | \subseteq | B |$. We expand B to a structure B_A for $L(A)$ such that if i is the name of an individual a of A, then B_A assigns a to i.

The notion of *Γ-diagram* of A is the same as in the classical case, and denoted by $D_\Gamma(A)$. If Γ is the set of open formulas, we write $D(A)$ for $D_\Gamma(A)$; if Γ is the set of all formulas, we write $D_F(A)$ for $D_\Gamma(A)$.

Theorem 4.22 (Diagram lemma) *Let Γ be a set of formulas, and let A and B be structures of for L such that $| A | \subseteq | B |$. Then, A is a Γ-substructure of B iff B_A is a model of $D_\Gamma(A)$.*

Let Γ be a set of formulas in L. We say that Γ is *regular* if every formula of the form $x = y$ or $\neg(x = y)$ is in Γ, and if for every formula $A \in \Gamma$, every formula of the form $A(x_1, \ldots, x_n) \in \Gamma$.

We obtain two versions of the model extension theorem.

Theorem 4.23 (Model extension theorem I) *Let A be a structure for L, T a theory in the language L, and Γ a regular set of formulas in L. Then, A has a Γ-extension that is a model of T iff every theorem of T that is a disjunction of strong negations of formulas in Γ is valid in A.*

Proof Similar to the classical case, using the reduction theorem for trivialization I.

Theorem 4.24 (Model extension theorem II) *Let A be a structure for L, T a theory in the language L, and Γ a regular set of formulas in L. Then, A has a Γ-extension that is a model of T iff every theorem of T that is a disjunction of negations of complex formulas in Γ is valid in A.*

Proof Similar to the classical case, using the reduction theorem for trivialization II.

From Theorems 4.23 and 4.24, the following corollaries follow.

Corollary 4.5 *Let Γ be a regular set of formulas in L and let Δ be a set of formulas containing every formula $\forall x_1 \ldots \forall x_n A$, where A is a disjunction of strong negations of formulas in Γ. If A is a structure for L and B is a Δ-extension of A, then there is a Γ-extension of C of B that is an elementary extension of A.*

Corollary 4.6 *Let Γ be a regular set of formulas in L and let Δ be a set of formulas containing every formula $\forall x_1 \ldots \forall x_n A$, where A is a disjunction of negations of complex formulas in Γ. If A is a structure for L and B is a Δ-extension of A, then there is a Γ-extension of C of B that is an elementary extension of A.*

When Γ is either the set of open formulas or the set of existential formulas, the preceding results provide some conditions on the models of T that determine T is equivalent to a theory whose non-logical axioms are in Γ.

Theorem 4.25 *Let Γ be a set of formulas in L and Γ' be the set of formulas of L that are theorems of T. If every structure for $L(T)$ in which all the formulas of Γ' are valid is model of T, then T is equivalent to a theory whose non-logical axioms are n Γ.*

Theorem 4.26 (Łoś-Tarski theorem) *A theory T is equivalent to an open theory iff every substructure of a model of T is a model of T.*

Proof As in the corresponding classical case, implying our version of the model extension theorem (I or II).

A sequence $(A_i)_{i \in N^*}$ of structure for L is called a *chain* if for each n, A_{n+1} is an extension of A_n. Given a chain, we define a structure A called the *union* of the chain.

The universe of A is $\mid A \mid = \bigcup_{i=1}^{\infty} \mid A_i \mid$. If $a_1, \ldots, a_k \in \mid A \mid$, then there is an n such that all $a_1, \ldots, a_k \in \mid A \mid$. We then set $f_A(a_1, \ldots, a_k) = f_{A_n}(a_1, \ldots, a_k)$ and $p_A(a_1, \ldots, a_k) = \lambda$ iff $p_{A_n}(a_1, \ldots, a_k) = \lambda, \forall \lambda \in \mid \tau \mid$. It is immediate that this definition is well-defined.

An *elementary chain* is a chain $(A_i)_{i \in N^*}$ such that for each n, A_{n+1} is an elementary extension of A_n.

Theorem 4.27 (Tarski) *If $(A_i)_{i \in N^*}$ is an elementary chain, then the union $A = \bigcup_{i=1}^{\infty} A_i$ is an elementary extension of each A_i.*

Proof By induction on the length of A, where A is a closed formula in $L(A_i)$ in order to prove that $A_i(A) = A(A)$, using Corollary 4.4.

Theorem 4.28 (Chang-Łoś-Suszko theorem) *A theory is equivalent to a theory whose non-logical axioms are existential iff every union of a chain of models of T is a model of T.*

Proof The proof that the condition holds if T is equivalent to a theory of T, whose non-logical axioms are existential, is formally the same as in the classical case. So, suppose that every union of a chain of models of T is a model of T. We shall construct a chain $(A_i)_{i \in N^*}$ such that $A_1 = A$, A_{2i} is a model of T, and A_{2i+3} is an elementary extension of A_{2i+1}. To construct the chain, we use the model extension theorem (I or II). We need to establish that if A is a disjunction of strong negation of universal formulas, then A is equivalent to a formula B n prenex from, that is existential. But, this is true in virtue of the equivalences like $(A \wedge B) \leftrightarrow \neg_*(\neg_* A \vee \neg_* B)$, $\neg_* \forall x A \leftrightarrow \exists x \neg_* A$, $\exists x B \vee C \leftrightarrow \exists x(B \vee C)$.

Given such a chain, we let B be $\bigcup_{i=1}^{\infty} A_i$. Then, B is the union of the chain $(A_{2i})_{i \in N^*}$ of models of T, and hence, is a model of T. But, B is also the union of the elementary chain $(A_{2i+1})_{i \in N^*}$. Hence, by Tarski's theorem (Theorem 4.27), B is an elementary extension of $A_1 = A$, and therefore is elementarily equivalent to A. It follows that A is a model of T.

Theorem 4.29 *If a closed formula A is* $\forall\exists$*-formula valid in each structure of a chain* $(A_i)_{i \in N^*}$*, then A is valid in each union of the chain.*

The classical results on cardinality of models like the cardinality theorem of Tarski and Löwenheim-Skolem theorem, can be immediately extended, since they are theorems of a structural character.

Now, let T and T' be theories the *union* of T and T', denoted $T \cup T'$, is the theory whose non-logical symbols are the non-logical symbols of T', and whose non-logical axioms are all the non-logical axioms of both T and T'.

Theorem 4.30 (Craig-Robinson theorem I) *Let T and* T' *be theories. Then,* $T \cup T'$ *is trivial iff there is a closed formula A such that* $\vdash_T A$ *and* $\vdash_T \neg_* A$.

Proof Similar to the classical case, using the strong negation and the notion of trivial theory.

Theorem 4.31 (Craig-Robinson theorem II) *Let T and* T' *be theories. Then,* $T \cup T'$ *is trivial iff there is a closed complex formula A such that* $\vdash_T A$ *and* $\vdash_T \neg A$.

Proof Similar to the classical case, using the negation of complex formula and the notion of trivial theory.

Then, Craig interpolation theorem follows.

Theorem 4.32 (Craig interpolation theorem) *Let T and* T' *be theories, and let A and B formulas such that* $\vdash_{T \cup T'} A \to B$*, where A is a formula T and B is a formula of* T'*. Then, there is a formula C such that* $\vdash_T A \to C$ *and* $\vdash_{T'} C \to B$.

We can also define the definability of a symbol in terms of a set of symbols. By using the concept of u-isomorphism, we have the following Beth's definability theorem by the application of the interpolation theorem.

Theorem 4.33 (Beth's definability theorem) *Let Q be a set of non-logical symbols in T, and u be a non-logical symbol of T that is not in Q. Then, u is definable in terms of Q in T iff $\varphi :| A | \rightarrow | B |$ that is a v-morphism for every $v \in Q$, φ is a u-isomorphism.*

All standard results in classical model theory can be extended for annotated model theory.

The ultraproduct is an important concept in model theory; see Bell and Samson [48]. Abe and Akama investigated the ultraproduct method for $Q\tau$ in [7]. Naturally, other important basic topics, such as applications of complete theories (e.g. some version of the quantifier elimination theorem) and saturated model theory remain to be studied.

4.4 Proof Methods

We have already given an axiomatization of annotated logics by a Hilbert system. Unfortunately, Hilbert systems are not suited for practical purposes. For this reason, it is necessary to work out proof methods for annotated logics. Possible methods include tableau, natural deduction, and resolution. Below we present a natural deduction system. We will discuss resolution in connection with logic programming in Chap. 6.

Now, we describe a proof theory based on *natural deduction* for annotated logics. A natural deduction formulation for $Q\tau$ was explored in Akama et al. [31].

Natural deduction was invented by Gentzen to formalize human reasoning in 1934. He also proposed a *sequent calculus* for technical purposes; see [85]. Gentzen provided natural deduction systems (also sequent calculi) for classical and intuitionistic logics. Later, it was polished by Prawitz [135].

Natural deduction is formalized by a set of rules. Generally, it has no axioms but some formulations use axioms. A *rule* in natural deduction is of the form:

$$\frac{A}{B}$$

where A denotes *premises* and B a *consequent*, respectively.

There are two types of rules: *introduction rule* and *elimination rule*. An introduction rule introduces a logical symbol in the consequent, and an elimination rule eliminates a logical symbol in the consequent. If B is derived from the assumption A, we write this as $\substack{[A] \\ B}$. A *proof* is constructed as a tree whose root is a formula to be proved, and all assumptions must be *discharged* in that they do not occur in the consequent after the application of a rule.

We denote by $NQ\tau$ a natural deduction system for $Q\tau$ which is formalized by an axiom and a set of rules. We use $\vdash_{NQ\tau} A$ to mean that A is provable in $NQ\tau$. The rules for $NQ\tau$ consist of rules for classical predicate logic and rules for annotation.

Natural Deduction System $NQ\tau$

(Axiom)

$$(\tau_1)\frac{}{P_\bot(t_1, \ldots, t_n)}$$

(Rules)

$$(\wedge I)\frac{A \quad B}{A \wedge B} \qquad (\wedge E)\frac{A \wedge B}{A} \quad \frac{A \wedge B}{B}$$

$$(\vee I)\frac{A}{A \vee B} \quad \frac{B}{A \vee B} \qquad (\vee E)\frac{A \vee B \quad \begin{array}{c}[A]\\ C\end{array} \quad \begin{array}{c}[B]\\ C\end{array}}{C}$$

$$(\rightarrow I)\frac{\begin{array}{c}[A]\\ B\end{array}}{A \rightarrow B} \qquad (\rightarrow E)\frac{A \quad A \rightarrow B}{B}$$

$$(P)\frac{\begin{array}{c}[A \rightarrow B]\\ A\end{array}}{A} \qquad (\neg I)\frac{\begin{array}{c}[F]\\ G \wedge \neg G\end{array}}{\neg F}$$

$$(\neg\neg E)\frac{\neg\neg F}{F} \qquad (\neg E)\frac{F \quad \neg F}{A}$$

$$(\forall I)\frac{A(a)}{\forall x A(x)} \qquad (\forall E)\frac{\forall x A(x)}{A(t)}$$

$$(\exists I)\frac{A(t)}{\exists x A(x)} \qquad (\exists E)\frac{\exists x A(x) \quad \begin{array}{c}[A(a)]\\ B\end{array}}{B}$$

$$(\tau_2)\frac{\neg^k P_\lambda(t_1, \ldots, t_n)}{\neg^{k-1} P_{\sim\lambda}(t_1, \ldots, t_n)} \quad \frac{\neg^{k-1} P_{\sim\lambda}(t_1, \ldots, t_n)}{\neg^k P_\lambda(t_1, \ldots, t_n)}$$

$$(\tau_3)\frac{P_\lambda(t_1, \ldots, t_n) \quad \lambda \geq \mu}{P_\mu(t_1, \ldots, t_n)} \qquad (\tau_4)\frac{P_{\lambda_1}(t_1, \ldots, t_n) \ldots P_{\lambda_m}(t_1, \ldots, t_n) \; \lambda = \bigvee_{i=1}^{m} \lambda_i}{P_\lambda(t_1, \ldots, t_n)}$$

Here, in the quantifier rules t is an arbitrary term and a is a term subject to the variable condition. For the non-atomic level, $Q\tau$ is positive classical predicate logic. Thus, we need (P) a la Curry; see Curry [59]. In place of $(\neg\neg E)$, we can add the axiom of the form: $F \vee \neg F$.

It is not difficult to show the equivalence of Hilbert and natural deduction versions of $Q\tau$. Here, we give a completeness proof of $Q\tau$ using the natural calculus $NQ\tau$. Although it is possible to have a purely syntactic proof, we here employ the semantics given in Chap. 3.

Theorem 4.34 shows the soundness of $NQ\tau$.

Theorem 4.34 (Soundness) *If* $\vdash_{NQ\tau} A$, *then* $\models_{NQ\tau} A$.

Proof It suffices to check that the axiom is valid and that each rule preserves validity. For classical rules, proofs are straightforward. The axiom (τ_1) is trivially valid via the semantics.

We need to check four rules. For (P): Suppose that $\models (A \to B) \to A$ and that $\not\models A$. From the former, we have that $\not\models A \to B$. This implies that $\models A$ and $\not\models B$. But, $\models A$ contradicts the assumption. Thus, we can conclude that there is no interpretation I such that $I \models (A \to B) \to A$ and $I \not\models A$.

For (τ_2): Trivial by definition of a hyper-literal.

For (τ_3): Immediate from the property of lattice ordering.

For (τ_4): Suppose that $\models P_{\lambda_1}(t_1, \ldots, t_n), \ldots, \models P_{\lambda_m}(t_1, \ldots, t_n)$ and $\lambda = \bigvee_{i=1}^{m} \lambda_i$ hold but $\not\models P_\lambda(t_1, \ldots, t_n)$. Now, pick up the supremum annotation λ_i $(1 \leq i \leq n)$. Then, $\lambda = \lambda_i$. This implies that $\not\models P_{\lambda_i}(t_1, \ldots, t_n)$, contradicting the initial assumption.

For completeness, we use the concept of $Q\tau$ *-saturated set* of formulas. We say that a set of formulas Γ is *maximal non-trivial* iff it is trivial and $A \in \Gamma$ or $\neg A \in \Gamma$ for every formula A.

Definition 4.31 (*$Q\tau$-saturated set*) A *$Q\tau$-saturated set* is a set of formulas Γ satisfying the following:

(i) Γ is maximal non-trivial.
(ii) If $\Gamma \vdash A$ then $A \in \Gamma$.
(iii) If $\exists x A(x) \in \Gamma$, then $A(a) \in \Gamma$ for some constant a.

Theorem 4.35 *A non-trivial set of formulas Γ can be extended to the $Q\tau$-saturated set of formulas with respect to the set of all formulas $Form$ in $Q\tau$.*

Proof We construct the sequence $\Gamma_0 = \Gamma, \Gamma_1, \ldots, \Gamma_n, \ldots$. Here, we expand the language L with the set C of constants new in L and set $L' = L \cup C$. Let $A_1, A_n, \ldots$ be an enumeration of formulas of $Q\tau$. First, we check (i) in Definition 4.31. Consider the case that n is even. Then, Γ_{i+1} is defined as follows:

$$\Gamma_{i+1} = \begin{cases} \Gamma \cup \{A_{i+1}\} & \text{if } \Gamma_i \cup \{A_{i+1}\} \text{ is non-trivial} \\ \Gamma_i & \text{otherwise} \end{cases}$$

Here, each set of the sequence $\Gamma_0, \Gamma_1, \ldots$ is non-trivial and this is a non-decreasing sequence of sets such that $\Gamma_0 \subseteq \Gamma_1 \subseteq \ldots$. Then, we set:

$$\Gamma^* = \bigcup_{i=0}^{\infty} \Gamma_i$$

We can see that Γ^* is a maximally non-trivial set containing Γ. In fact, each finite subset of Γ^* must be non-trivial since Γ_i is non-trivial. It follows that Γ^* itself is non-trivial.

Now, we claim that (i) Γ^* is a maximal non-trivial set. Now, suppose $A \notin \Gamma^*$ and $\neg A \notin \Gamma^*$. As A is a formula in $Q\tau$, it must appear in the above enumeration, say as A_k. If $\Gamma \cup \{A_k\}$ were non-trivial, then our construction would guarantee that $A_{k+1} \in \Gamma_{k+1}$, and hence that $A_k \in \Gamma^*$. From $A_k \notin \Gamma^*$, we have that $\Gamma_k \cup \{A_k\}$ is trivial. Thus, $\Gamma^* \cup \{A_k\}$ is trivial. It follows that Γ^* is a maximal non-trivial set. As $\Gamma \subseteq \Gamma_i$ $(i \in \omega)$, the following holds.

$$\Gamma \subseteq \Gamma^* = \bigcup_{i=0}^{\infty} \Gamma_i$$

On the other hand, suppose that Γ^* is trivial. Then, $\Gamma^* = Form$ follows. Thus, we have that $P_\lambda(t_1, \ldots, t_n), \neg_* p_\lambda(t_1, \ldots, t_n) \in \Gamma^*$. As $\mid \tau \mid$ is finite, any application of natural deduction rules has only a finite number of premises. Thus, there are $n, m < \omega$ such that $P_\lambda(t_1, \ldots, t_n) \in \Gamma_n$ and $\neg_* P_\lambda(t_1, \ldots, t_n) \in \Gamma_m$. Therefore, $P_\lambda(t_1, \ldots, t_n), \neg_* P_\lambda(t_1, \ldots, t_n) \in \Gamma_{n_0}$, where $n_0 = max(n, m)$. Thus, Γ_{n_0} is trivial and we reach a contradiction.

(ii) is straightforward by $(\rightarrow I)$ and maximality of Γ^*.

For (iii), consider the enumeration in which n is odd. Let $\exists x A(x)$ be the first formula of the enumeration of existential formulas not considered before satisfying $\Gamma_n \vdash \exists x A(x)$. If we assume that a is a constant not in Γ_n, then $\Gamma_{n+1} = \Gamma_n \cup \{A(a)\}$. Otherwise $\Gamma_{n+1} = \Gamma_n$. Now, suppose $\exists x A(x) \in \Gamma^*, A(a) \notin \Gamma^*$. By construction, $\Gamma^* \vdash A(a) \rightarrow B$. By $(\exists E)$, $\Gamma^* \vdash \exists x A(x) \rightarrow B$. A contradiction. Thus, (iii) is verified. It is therefore shown that Γ^* is $Q\tau$-saturated.

Lemma 4.2 *Let Γ be a $Q\tau$-saturated set of formulas. Then, the following hold.*

(1) $A \wedge B \in \Gamma$ *iff* $A \in \Gamma$ *and* $B \in \Gamma$.
(2) $A \vee B \in \Gamma$ *iff* $A \in \Gamma$ *or* $B \in \Gamma$.
(3) $A \rightarrow B \in \Gamma$ *iff* $A \notin \Gamma$ *and* $B \in \Gamma$.
(4) $\neg_* A \in \Gamma$ *iff* $A \notin \Gamma$.
(5) $\neg^k P_\lambda(t_1, \ldots, t_n) \in \Gamma$ *iff* $\neg^{k-1} P_{\sim\lambda}(t_1, \ldots, t_n) \in \Gamma$.
(6) *If* $P_\lambda(t_1, \ldots, t_n), P_\mu(t_1, \ldots, t_n) \in \Gamma$, *then* $P_\theta(t_1, \ldots, t_n) \in \Gamma$, *where* $\theta = max(\lambda, \mu)$.
(7) $\exists x A(x) \in \Gamma$ *iff* $A(a) \in \Gamma$ *for some* a.
(8) $\forall x A(x) \in \Gamma$ *iff* $A(a) \in \Gamma$ *for all* a.

Proof (1)–(4) are the same as classical cases. (5) is trivial by the definition of $Q\tau$-saturated set. For (6), suppose $P_\lambda(t_1, \ldots, t_n)$, $P_\mu(t_1, \ldots, t_n) \in \Gamma$. By (1), we have $P_\lambda(t_1, \ldots, t_n) \wedge P_\mu(t_1, \ldots, t_n) \in \Gamma$. But, by ($\tau 4$), $P_\theta(t_1, \ldots, t_n) \in \Gamma$, where $\theta = max(\lambda, \mu)$. For (7), one direction is immediate from the definition of $Q\tau$-saturated set. The other direction can be obtained by ($\exists I$). (8) is treated similarly. We finish by checking the $Q\tau$-saturatedness.

Definition 4.32 (*Canonical interpretation*) Let the language of *FN* be L, L' be $L \cup C'$, and Var_L and C_L be the set of variables and of constants in L. Then, a canonical interpretation I^* for a $Q\tau$-saturated set Γ of formulas is defined as follows:

1. $dom(I^*) = C_{L'}$
2. $\alpha_{I^*}, \beta_{I^*}, \gamma_{I^*}$ are defined above.
3. The variable assignment v^* is a function from $Var_{L'}$ to $dom(I^*)$.
4. $d_{I^*,v^*}(t)$ is defined above.
5. χ_Γ is a characteristic function from $Form$ to $\mathbf{2} = \{0, 1\}$.

Let $Atom$ be a set of annotated atoms. Then, we define the function $I_*^{v_*}$ from $Atom$ to $|\tau|$ satisfying $I_*^{v_*}(P(t_1, \ldots, t_n)) = \bigvee\{\mu \in |\tau| : P_\mu(t_1, \ldots, t_n) \in \Gamma\}$. Next, we define the function $Val_{I_*}^{v_*}$ from $Form$ to $\mathbf{2}$. It suffices to show that $Val_{I_*}^{v_*} = \chi_\Gamma$. Assume $P_\mu(t_1, \ldots, t_n) \in \Gamma$. Then, we have that $\chi_\Gamma(P_\mu(t_1, \ldots, t_n)) = 1$. On the other hand, it is clear that $\gamma_{I_*}(P)(d_{I_*,v_*}(t_1), \ldots, d_{I_*,v_*}(t_n)) \geq \mu$. So, $Val_{I_*}^{v_*}(P_\mu(t_1, \ldots, t_n)) = 1$. If $P_\mu(t_1, \ldots, t_n) \notin \Gamma$, then $\chi_\Gamma(P_\mu(t_1, \ldots, t_n)) = 0$. It is also not the case that $\gamma_{I_*}(P)(d_{I_*,v_*}(t_1), \ldots, d_{I_*,v_*}(t_n)) \geq \mu$, we have that $P_{\gamma_{I_*}(P)(d_{I_*,v_*}(t_1),\ldots,d_{I_*,v_*}(t_n))}(t_1, \ldots, t_n) \in \Gamma$, which is a contradiction. Thus, it is not the case that $\gamma_{I_*}(P)(d_{I_*,v_*}(t_1), \ldots, d_{I_*,v_*}(t_n)) \geq \mu$. Hence, we conclude that $Val_{I_*}^{v_*}(P_\mu(t_1, \ldots, t_n)) = 0$.

By Theorem 2.15(5), $\neg^k P_\mu(t_1, \ldots, t_n) \in \Gamma$ iff $\neg^{k-1} P_{\sim\mu}(t_1, \ldots, t_n) \in \Gamma$. Thus, $\chi_\Gamma(\neg^k P_\mu(t_1, \ldots, t_n)) = \chi_\Gamma(\neg^{k-1} P_{\sim\mu}(t_1, \ldots, t_n))$ holds.

We have to prove that $Val_{I_*}^{v_*}(\neg^k P_\mu(t_1, \ldots, t_n)) = \chi_\Gamma(\neg^k P_\mu(t_1, \ldots, t_n))$ by induction on k. If $k = 0$, then the proposition is obvious. Here, assume that it holds for $k - 1 (k \geq 1)$. Then, $\chi_\Gamma(\neg^k P_\mu(t_1, \ldots, t_n)) = \chi_\Gamma(\neg^{k-1} P_{\sim\mu}(t_1, \ldots, t_n)) = Val_{I_*}^{v_*}(\neg^{k-1} P_{\sim\mu}(t_1, \ldots, t_n)) = Val_{I_*}^{v_*}(\neg^k P_\mu(t_1, \ldots, t_n))$.

Theorem 4.36 *Let Γ be $Q\tau$-saturated set of formulas. Then, we have:*

$\chi_\Gamma(A) = 1$ *iff* $A \in \Gamma$ *and*
$\chi_\Gamma(A) = 0$ *iff* $A \notin \Gamma$.

Proof For annotated atom and hyper-literal, it is clear from the above definition. The cases that A is any formula can be easily proved. Assume that if A is of the form $\neg B$, where B is a complex formula. So, $\chi_\Gamma(B) = Val_{I^*}^{v_*}(B)$. If $A \in \Gamma$, then $B \notin \Gamma$.

Then, we have that $\chi_{\Gamma}(A) = 0$ and $\chi_{\Gamma}(B) = 1$. But, $Val_{I_*}^{v_*}(A) = 1 - Val_{I_*}^{v_*}(B)$. Thus, $Val_{I_*}^{v_*}(A) = 0$ follows. The other cases present no difficulty.

From Theorem 4.36, completeness follows.

Theorem 4.37 (Completeness) *If* $\models_{NQ\tau} A$, *then* $\vdash_{NQ\tau} A$.

Theorem 4.37 is regarded as another version of the completeness theorem of $Q\tau$. It is easy to extend it for strong completeness. It is interesting to prove syntactic proof of completeness by normal form theorem by using the technique of Prawitz [135].

We can also work out a sequent calculus for $Q\tau$, but no work has been done so far. In particular, it is of importance to prove the cut-elimination theorem because the subject is closely related to completeness and automated deduction.

Smullyan [148] developed *tableau calculus* in detail. The idea can be applied to many non-classical logics. Akama et al. [29] presented a tableau formulation of $P\tau$. It can be expanded for $Q\tau$. As there are intimate connections of sequent and tableau calculi, a desired sequential formulation could be obtained.

4.5 Annotated Modal Logics

Modal logic extends classical logic with modal operators to express intensional concepts like necessity and possibility. Modal logic has many applications in various areas including philosophy, linguistics and computer science. The reader is referred to Hughes and Cresswell [91, 92].

It is interesting to present *annotated modal logic* in order to deal with intensional concepts. This can be done by extending annotated logics with modal operators. Annotated modal logics $S5\tau$ were proposed by Abe [2], and later generalized by Akama and Abe [23]. Here, we set up annotated modal logics $K\tau$, which correspond to an annotated version of classical modal logic K.

Definition 4.33 (*Symbols*) The symbols of $K\tau$ are defined as follows:

1. Propositional symbols: $p, q, \ldots$ (possibly with subscript)
2. Annotated constants: $\mu, \lambda, \ldots \in |\tau|$
3. Logical connectives: $\wedge$ (conjunction), $\vee$ (disjunction), $\rightarrow$ (implication), and $\neg$ (negation)
4. Modal operator: $\Box$ (necessity)
5. Parentheses: (and)

Non-modal formulas of $K\tau$ are defined by Definition 2.2. In addition, we define a modal sentence as follows: If A is a formula and $\Box$ is the *necessity operator*, then $\Box A$ is a formula. $\Box A$ reads "A is necessary".

Let $\diamond$ be the *possibility operator*. A formula of the form $\diamond A$ is introduced by definition as follows:

$$\diamond A =_{def} \neg_* \Box \neg_* A$$

Here, $\neg_*$ denotes strong negation. $\diamond A$ reads "A is possible". If modal operators are read differently, we can formalize various modal logics such as temporal logics and epistemic logics.

Now, we present a Hilbert style axiomatization of $K\tau$. Let A, B, C be arbitrary formulas, F, G be complex formulas, p be a propositional variable, and λ, μ, λ_i be an annotated constant. Then, the postulates are as follows:

Postulates for $K\tau$

$(\rightarrow_1)$ $(A \rightarrow (B \rightarrow A)$
$(\rightarrow_2)$ $(A \rightarrow (B \rightarrow C)) \rightarrow ((A \rightarrow B) \rightarrow (A \rightarrow C))$
$(\rightarrow_3)$ $((A \rightarrow B) \rightarrow A) \rightarrow A$
$(\rightarrow_4)$ $A, A \rightarrow B/B$
$(\wedge_1)$ $(A \wedge B) \rightarrow A$
$(\wedge_2)$ $(A \wedge B) \rightarrow B$
$(\wedge_3)$ $A \rightarrow (B \rightarrow (A \wedge B))$
$(\vee_1)$ $A \rightarrow (A \vee B)$
$(\vee_2)$ $B \rightarrow (A \vee B)$
$(\vee_3)$ $(A \rightarrow C) \rightarrow ((B \rightarrow C) \rightarrow ((A \vee B) \rightarrow C))$
$(\neg_1)$ $(F \rightarrow G) \rightarrow ((F \rightarrow \neg G) \rightarrow \neg F)$
$(\neg_2)$ $F \rightarrow (\neg F \rightarrow A)$
$(\neg_3)$ $F \vee \neg F$
(τ_1) $p_\perp$
(τ_2) $\neg^k p_\lambda \leftrightarrow \neg^{k-1} p_{\sim\lambda}$
(τ_3) $p_\lambda \rightarrow p_\mu$, where $\lambda \geq \mu$
$(\tau_4)]$ $p_{\lambda_1} \wedge p_{\lambda_2} \wedge \ldots \wedge p_{\lambda_m} \rightarrow p_\lambda$, where $\lambda = \bigvee_{i=1}^{m} \lambda_i$
$(\Box_1)$ $\Box(A \rightarrow B) \rightarrow \Box(A \rightarrow B)$
$(\Box_2)$ $A/\Box A$

Here, $(\Box_1)$ denotes the distribution axiom (K) and $(\Box_2)$ the rule of inference, called the *necessitation* (NEC). $K\tau$ are the minimal normal annotated logics. The consequence relation $\vdash_{K\tau}$ is defined as usual.

As in classical modal logic, we can axiomatize various annotated modal logics by adding axioms to $K\tau$. The well known axioms are as follows:

$(\Box_T)$ $\Box A \rightarrow A$
$(\Box_D)$ $\Box A \rightarrow \diamond A$
$(\Box_4)$ $\Box A \rightarrow \Box\Box A$
$(\Box_5)$ $\diamond A \rightarrow \Box\diamond A$

Annotated modal logics $T\tau$ are obtained by $K\tau$ with $(\Box_T)$, $S4\tau$ by $T\tau$ with $(\Box_4)$, and $S5\tau$ with $(\Box_5)$.

We turn to *Kripke semantics* for annotated modal logics. A Kripke semantics can be developed by means of a *Kripke model* based on the concept of *possible world*.

Definition 4.34 (*Kripke model*) A Kripke model for $K\tau$ is a tuple $\mathcal{M} = (W, R, V)$ where

1. W is a non-empty set of possible worlds.
2. R is a binary relation on W.
3. V is a valuation function $W \times \mathbf{P} \rightarrow |\tau|$.

Here, $\mathcal{F} = (W, R)$ is called the *frame*. If $w \in W$, $p \in \mathbf{P}$, and $\lambda \in |\tau|$ are such that $V(w, p) \geq \lambda$, then w say that p_λ is true in a world w in a model $\mathcal{M}$, and false otherwise. We can then define the truth relation $\mathcal{M}, w \models A$ (A is true at w in $\mathcal{M}$) for any formula A.

Definition 4.35 (*Truth definition*) The truth relation $\mathcal{M}, w \models A$ is defined as follows:

1. If $p \in \mathbf{P}$ and $\lambda \in |\tau|$ then
   a. $\mathcal{M}, w \models p_\lambda \Leftrightarrow V(w, p) \geq \lambda$
   b. $\mathcal{M}, w \models \neg^k p_\lambda \Leftrightarrow \mathcal{M}, w \models \neg^{k-1} p_{\sim\lambda}$ $(k > 0)$
2. If A and B are formulas, then
   a. $\mathcal{M}, w \models A \wedge B \Leftrightarrow \mathcal{M}, w \models A$ and $\mathcal{M}, w \models B$
   b. $\mathcal{M}, w \models A \vee B \Leftrightarrow \mathcal{M}, w \models A$ or $\mathcal{M}, w \models B$
   c. $\mathcal{M}, w \models A \rightarrow B \Leftrightarrow \mathcal{M}, w \not\models A$ or $\mathcal{M}, w \models B$
   d. $\mathcal{M}, w \models \Box A \Leftrightarrow \forall v(wRv \Rightarrow \mathcal{M}, v \models A)$
3. If F is a complex formula,

   $\mathcal{M}, w \models \neg F \Leftrightarrow \mathcal{M}, w \not\models F$

Here, R has no condition. Let $\mathcal{M}$ be a Kripke model. Then, we say that A is true in $\mathcal{M}$ iff $\mathcal{M}, w \models A$ for any $w \in W$. A formula A is called *valid*, written $\models_{K\tau} A$, iff $\mathcal{M}, w \models A$ for any $w \in W$ and for any $\mathcal{M}$. A semantic consequence relation $\Gamma \models A$ is also defined in such a way that for any $\mathcal{M}$ and any $B \in \Gamma$ if $\models_{K\tau} B$ then $\models_{K\tau} B$.

There are conditions on R satisfying extensions of $K\tau$. For annotated modal logics $T\tau$, R must be the reflexive relation. For annotated modal logics $D\tau$, R must be the serial relation. For $S4\tau$, R must be the reflexive and transitive relation, and for $S5\tau$, R must be the equivalence relation. We chose an appropriate relation R for our purposes. For $S5\tau$, the truth relation for $\Box$ is simplified without reference to R as follows:

$$\mathcal{M}, w \models \Box A \Leftrightarrow \forall v(\mathcal{M}, v \models A)$$

The notions of truth, validity and semantic consequence of various annotated modal logics are the same as in annotated modal logics $K\tau$ defined above.

The basic results of $P\tau$ shown in Chap. 2 also hold for $K\tau$.

Lemma 4.3 *Let p be a propositional variable and* $\mu, \lambda \in |\tau|$. *Then, we have:*

1. $\models p_{\perp}$
2. $\models p_{\mu} \rightarrow p_{\lambda},\ \mu \geq \lambda$
3. $\models \neg^{k} p_{\mu} \leftrightarrow p_{\sim^{k}\mu},\ k \geq 0$

Lemma 4.4 *For any formula A of* $K\tau$, *we do not have simultaneously*

$\models_{K\tau} A$ *and* $\models_{K\tau} \neg_{*} A$.

Theorem 4.38 *There are Kripke models* $\mathcal{M} = (W, R, V)$ *such that for some hyper-literal A and B and some worlds* $w, w' \in W$, *we have that* $\mathcal{M}, w \models A$ *and* $\mathcal{M}, w \models \neg A$ *but* $\mathcal{M}, w' \not\models B$.

Theorem 4.39 *There are Kripke models* $\mathcal{M} = (W, R, V)$ *such that for some hyper-literal A and B and some world* $w \in W$, *we have that neither* $\mathcal{M}, w \models A$ *nor* $\mathcal{M}, w \models \neg A$.

In $K\tau$, there are inconsistent worlds, paracomplete worlds, or both. These worlds are interpreted as *non-classical worlds*. We now extend the notions of paraconsistency and paracompleteness for Kripke models.

Definition 4.36 A Kripke model $\mathcal{M} = (W, R, V)$ is called *paraconsistent* iff there are atomic formulas p and q and annotated constants $\lambda, \mu \in |\tau|$ such that $\mathcal{M}, w \models p_{\lambda}$ and $\mathcal{M}, w \models \neg p_{\lambda}$ but $\mathcal{M}, w \not\models q_{\mu}$.

Definition 4.37 A system $K\tau$ is called *paraconsistent* iff there is a Kripke model $\mathcal{M} = (W, R, V)$ for $K\tau$ such that $\mathcal{M}$ s paraconsistent.

Theorem 4.40 $K\tau$ *is paraconsistent iff* $card(\tau) \geq 2$, *where* $card(\tau)$ *denotes the cardinality (cardinal number) of the set* τ.

Definition 4.38 A Kripke model $\mathcal{M} = (W, R, V)$ is called *paracomplete* iff there is an atomic formula p and annotated constants $\lambda \in |\tau|$ such that $\mathcal{M}, w \not\models p_{\lambda}$ and $\mathcal{M}, w \not\models \neg p_{\lambda}$.

Definition 4.39 A system $K\tau$ is called *paracomplete* iff there is a Kripke model $\mathcal{M} = (W, R, V)$ for $K\tau$ such that $\mathcal{M}$ is paracomplete.

Theorem 4.41 $K\tau$ *is paracomplete iff* $card(\tau) \geq 2$, *where* $card(\tau)$ *denotes the cardinality (cardinal number) of the set* τ.

Definition 4.40 A Kripke model $\mathcal{M} = (W, R, V)$ is called *non-alethic* iff is both paraconsistent and paracomplete.

Definition 4.41 A system $K\tau$ is called *non-alethic* iff there is a Kripke model $\mathcal{M} = (W, R, V)$ for $K\tau$ such that $\mathcal{M}$ is non-alethic.

Now, we prove soundness and completeness of $K\tau$. Theorem 4.42 shows the soundness of $K\tau$.

Theorem 4.42 (Soundness) *Let Γ be a set of formulas and A be any formula of $K\tau$. Then, we have that if $\Gamma \vdash_{K\tau} A$ then $\Gamma \models_{K\tau} A$.*

Proof The soundness of $P\tau$ was shown in Theorem 2.13. It thus suffices to prove that $(\Box_1)$ and $(\Box_2)$ are valid. For $(\Box_1)$, assume that $\Box(A \rightarrow B)$ is valid but $\Box A \rightarrow \Box B$ is not valid. Then, there is a Kripke model $\mathcal{M} = (W, R, V)$ and a world w such that $\mathcal{M}, w \models \Box(A \rightarrow B)$ and $\mathcal{M}, w \not\models \Box A \rightarrow \Box B$. From the former, we have that $\forall v(wRv \Rightarrow (\mathcal{M}v \models A \Rightarrow \mathcal{M}v \models B))$. From the latter we have that $\exists v(wRv$ and$\mathcal{M}, v \models A$ and $\mathcal{M}, v \not\models B)$. From these, we have that $\mathcal{M}, v \models B$ and $\mathcal{M}, v \not\models B$. A contradiction. Thus, $(\Box_1)$ is valid.

For $(\Box_2)$, assume that A is valid but $\Box A$ is not valid. Then, there is a Kripke mode $\mathcal{M} = (W, R, V)$ and a world w such that $\mathcal{M}, w \models A$ and $\mathcal{M}, w \not\models \Box A$. From the latter, we have that $\exists v(wRv$ and $\mathcal{M}, v \not\models A)$. Here, set $w = v$, producing a contradiction. Thus, $(\Box_2)$ is valid.

Next, we prove the completeness of $K\tau$. As in $P\tau$, we employ a maximal non-trivial set whose construction is shown in the proof of Theorem 2.14. It satisfies Theorem 4.43.

Theorem 4.43 *Let Γ be a maximal non-trivial subset of the set of formulas of $K\tau$. Let A and B be formulas. Then, we have the following:*

1. *If A is an axiom of $K\tau$, then $A \in \Gamma$.*
2. *$A, B \in \Gamma$ iff $A \wedge B \in \Gamma$.*
3. *$A \vee B \in \Gamma$ iff $A \in \Gamma$ or $B \in \Gamma$.*
4. *if $p_\lambda, p_\mu \in \Gamma$, then $p_\theta \in \Gamma$, where $\theta = max(\lambda, \mu)$.*
5. *$\neg^k p_\mu \in \Gamma$ iff $\neg^{k-1} p_{\sim\mu} \in \Gamma$, where $k \geq 1$.*
6. *if $A, A \rightarrow B \in \Gamma$, then $B \in \Gamma$.*
7. *$A \rightarrow B \in \Gamma$ iff $A \notin \Gamma$ or $B \in \Gamma$.*
8. *$A \in \Gamma$ iff $\neg_* A \notin \Gamma$. Moreover $A \in \Gamma$ or $\neg_* A \in \Gamma$.*
9. *If A is a complex formula, then $A \in \Gamma$ iff $\neg A \notin \Gamma$. Moreover $A \in \Gamma$ or $\neg A \in \Gamma$.*
10. *If $A \in \Gamma$, then $\Box A \in \Gamma$.*

Proof From the proof of Theorem 2.15, we prove (8), (9) and (10). For (8), it is immediate from $\vdash_{K\tau} A \vee \neg_* A$ (Theorem 2.9(1)).

For (9), it is immediate from $(\neg_3)$.

For (10), it is proved using $(\Box_2)$.

Given a set of Γ, we define $\Gamma_\Box = \{A \mid \Gamma \subseteq \Gamma\}$. Then, we can define a *canonical Kripke model*.

Definition 4.42 (*Canonical Kripke model*) A *canonical Kripke model* $\mathcal{M}'$ is a tuple (W^c, R^c, V^c) where

1. W^c is a maximal non-trivial set of formulas.
2. $R^c = \{(\Gamma, \Gamma') \mid \Gamma_\Box \subseteq \Gamma'\}$ for any $\Gamma, \Gamma' \in W^c$.
3. V^c is a valuation function $W^c \times \mathbf{P} \rightarrow \mid \tau \mid$, where $V^c(\Gamma, p) = \{\mu \in \mid \tau \mid \mid p_\mu \in \Gamma\}$.

From Theorem 2.15(4), for any $\Gamma \in W^c$, if $p_\lambda, p_\mu \in \Gamma$ then $p_\theta \in \Gamma$, where $\theta = max(\lambda, \mu)$. Then, $\mathcal{M}'$ is shown to be a well defined Kripke model.

Lemma 4.5 *Let p be a propositional variable. If Γ be a maximal non-trivial set of formulas. Then, we have that $p_{I(\Gamma,p)} \in \Gamma$.*

Proof Simple consequence of Theorem 4.43(4).

Lemma 4.6 *Let $\mathcal{M}^c = (W^c, R^c, V^c)$ be a canonical Kripke model, $\Gamma \in W^c$ and A a formula. Then, we have:*

$$\mathcal{M}', \Gamma \models A \Leftrightarrow A \in \Gamma$$

Proof As in the classical case; see Hughes and Cresswell [91, 92]. For an annotated atom, suppose $\mathcal{M}', \Gamma \models p_\lambda$. From Lemma 4.5, we have that $p_{V^c(\Gamma,p)} \in \Gamma$. It also follows that $V^c(\Gamma, p) \geq \lambda$. Since $p_{I(\Gamma,p)} \rightarrow p_\lambda$ is an axiom, $p_\lambda \in \Gamma$.

Suppose $p_\lambda \in \Gamma$. By Lemma 4.5, $p_{V^c(\Gamma,p)} \in \Gamma$. Then, we have that $V(\Gamma, p) \geq \lambda$. Thus, by definition, $\mathcal{M}', \Gamma \models p_\lambda$. By Theorem 4.43(5), $\neg^k p_\lambda \in \Gamma \Leftrightarrow \neg^{k-1} p_{\sim\lambda} p$. Thus, by definition, $\mathcal{M}^c, \Gamma \models \neg^k p_\lambda \Leftrightarrow \mathcal{M}^c, \Gamma \models \neg^{k-1} p_{\sim\lambda}$. By induction on k, we have that $\mathcal{M}', \Gamma \models p_\lambda$.

We can then reach the completeness theorem of $K\tau$.

Theorem 4.44 *For any formula A of $K\tau$, $\models_{K\tau} \Leftrightarrow \vdash_{K\tau} A$.*

It can be shown that $K\tau$ is decidable.

Theorem 4.45 (Decidability) *If τ is finite, then $K\tau$ is decidable.*

Proof Let $\mathcal{M}' = (W, R, V)$ is a Kripke model for $K\tau$, and $\mid W \mid$ be the number of worlds $\sharp W$ added with the number $\sharp R$. We show that there is an algorithm that, given a finite model $\mathcal{M}$ with a world $w \in W$, and a formula A, we can determine whether $\mathcal{M}, w \models A$ in time $k \; |k| \; \sharp \; |\tau| \; \sharp Sf(A)$, where $Sf(A)$ denotes the set of all subformulas of A. Let $A_1, A_2, \ldots, A_m$ be the subformulas of A and $A_m = A$ if A_i is a subformula of A_j, where $i < j$. We can label each world in W with A_j or $\neg_* A_i$ for $i = 1, 2, \ldots, k$ by induction on k whether A_j is true or not in w in time $ck \mid W \mid$ for some constant c. If A_{k+1} is of the form $\Box A_j$, where $j < k+1$, we label a world w with $\Box A_j$ iff each world w' such that wRw' is labeled with A_j or $\neg_* A_j$. This step can be carried out in time $m \mid W \mid \sharp \mid \tau \mid$. Then, we have the decidability of $K\tau$.

We can establish the completeness of some extensions of $K\tau$ by showing extra conditions of R^c in canonical Kripke model $\mathcal{M}' = (W^c, R^c, V^c)$. The decidability of such extensions can be also proved without any difficulty.

By interpreting modal operators in annotated modal logics in various ways, it is possible to present many annotated logics for intensional concepts. For instance, Abe and Akama proposed annotated temporal logics $\Delta\tau$ in [9] which can serve as logics for reasoning about time with inconsistency. It would be interesting to work out annotated logics for intensional concepts like knowledge, belief, and obligation.

Chapter 5
Variants and Related Systems

Abstract This chapter reviews some variants of annotated logics and related systems in the literature. Variants include fuzzy annotated logics, possibilistic annotated logics, inductive annotated logics, and structural annotated logics. We also compare annotated logics with related systems such as Labelled Deductive Systems and General Logics. Finally, we review systems of paraconsistent logics.

5.1 Fuzzy Annotated Logics

There are several variants of annotated logics which extend or modify original versions to improve them. Akama and Abe [25] proposed *fuzzy annotated logics* and Akama and Abe [26] developed *possibilistic annotated logics*. Da Costa and Krause's [67] *inductive annotated logics* are annotated systems for inductive reasoning.

Here, we present fuzzy annotated logics which deal with so-called *approximate reasoning*. Approximate reasoning is one of the key points when constructing an intelligent system, since human reasoning involves reasoning about uncertainty. There are many frameworks for approximate reasoning, e.g. fuzzy logic, probabilistic logic, DS-theory. However, these theories face two difficulties. One is a profound theoretical background. The other is a method of automated reasoning.

Fuzzy annotated logics $FP\tau$ are considered as a generalization of annotated logic $P\tau$ capable of describing fuzzy reasoning. In $FP\tau$, we allow the annotation on any level of formula to represent the uncertainty of a formula, whereas in $P\tau$ the annotation is possible only on the atomic level. $FP\tau$ enhances the expressive power of $P\tau$. In addition, $FP\tau$ is closely related to the *fuzzy operator logic* of Liu et al. [112, 158].

Fuzzy logic was proposed by Zadeh [159] as a formalism for approximate reasoning. Unfortunately, some people claim that fuzzy has no theoretical foundations. It is also recognized that automated reasoning in fuzzy logic is difficult. Of course, the two obstacles are interconnected. However, a lot work has been done to address these issues in the literature. One of the pioneering works is Lee's [108] paper on the resolution principle for fuzzy logic. We believe that Liu et al.'s work [112, 158] on fuzzy operator logic addresses the theoretical side with the possibility of developing

J.M. Abe et al., *Introduction to Annotated Logics*,
Intelligent Systems Reference Library 88, DOI 10.1007/978-3-319-17912-4_5

automated theorem-proving. The present attempt is to generalize such a work further in such a way that fuzzy logic can be recognized as an instance of annotated logics.

Fuzzy annotated logics $FP\tau$ uses the lattice $\tau = \langle|\tau|, \leq\rangle$, where $|\tau|$ is the set of truth-values corresponding to the unit interval [0, 1].

Definition 5.1 (*Symbols*) The symbols of $FP\tau$ are defined as follows:

1. Propositional symbols: $p, q, \ldots$ (possibly with subscript)
2. Annotated constants: $\mu, \lambda, \ldots \in |\tau|$
3. Logical connectives: $\wedge$ (conjunction), $\vee$ (disjunction), and $\rightarrow$ (implication)
4. Parentheses: (and)

Note that $FP\tau$ has no negations, since an annotation acts as a generalized negation.

Definition 5.2 (*Formulas*) Formulas are defined as follows:

1. If p is an atomic formula and $\lambda \in [0, 1]$, then λp is a *fuzzy formula*.
2. If A and B are formulas and $\lambda \in [0, 1]$, then $\lambda : A \wedge B$, $\lambda : A \vee B$, and $\lambda : A \rightarrow B$ are also formulas.

Here, the annotation is a real number. The fuzzy formula λp reads "the uncertainty (or fuzziness) of p is at least λ". Any formulas in $FP\tau$ can be given an annotation unlike $P\tau$.

The semantics of $FP\tau$ is described as a modification of the semantics of $P\tau$. There are several interesting semantics of $FP\tau$, one of which is presented below. For this purpose, we have to use the notion of *threshold of certainty* α. Then, we can specify that λp is true iff the uncertainty of p is at least λ and is greater than α. In general, most formalisms for fuzzy logic employ the threshold of certainty as the value 0.5, but this is not always essential in $FP\tau$. In this sense, we can dispense with negation. This is because the annotation can express both positive and negative degree of certainty.

Definition 5.3 (*Interpretation*) The interpretation I and the valuation v are the same as in $P\tau$. We here assume that $\alpha \in [0, 1]$ is the threshold of certainty. If λp is a fuzzy atomic formula, then:

$v(\lambda p) = 1$ iff $I(p) \geq \lambda > \alpha$
$v(\lambda p) = 0$ iff $I(p) < \lambda < \alpha$
If A and B are formulas, then the interpretations of $v(A \wedge B)$, $v(A \vee B)$, and $v(A \rightarrow B)$ are as in $P\tau$.

For fuzzy formulas, we must use the following principles before the interpretation.

If $\lambda > \alpha$, then:
$\lambda : A \wedge B = \lambda : A \wedge \lambda : B$
$\lambda : A \vee B = \lambda : A \vee \lambda : B$

$$\lambda : A \to B = \lambda : A \to \lambda : B.$$

If $\lambda < \alpha$, then

$$\lambda : A \wedge B = \lambda : A \vee \lambda : B,$$
$$\lambda : A \vee B = \lambda : A \wedge \lambda : B,$$
$$\lambda : A \to B = (1 - \lambda) : A \wedge \lambda : B.$$

The latter case involves interpretation of negation similar to the de Morgan laws.

We say that a formulas A is *valid* iff $v(A) = 1$ for all valuations v. From the proposed semantics, we can point out that the following formulas are in general not valid (Consider the case in which $\lambda = \alpha = 0.5$).

$\lambda A \vee (1 - \lambda)A$,
$(1 - \lambda)(\lambda A \wedge (1 - \lambda)A)$.

In fact, the first is a general form of the law of excluded middle ($A \vee \neg A$) and the second is a general form of the law of non-contradiction $\neg(A \wedge \neg A)$. This implies that $FP\tau$ is *paraconsistent* (i.e. both A and $\neg A$ can be true) and *paracomplete* (i.e. neither A nor $\neg A$ can be true).

Now, we describe an axiomatization of $FP\tau$ whose postulates are slightly different from those of $P\tau$ in Chap. 2.

Postulates for $FP\tau$

The postulates of $FP\tau$ consist of some of the postulates of $P\tau$, i.e. $(\to_1)$ to $(\vee_3)$, (τ_1), (τ_3), (τ_4) and the following:

(FP1) $\lambda : A \wedge B \leftrightarrow \lambda : A \wedge \lambda : B$, where $\lambda > \alpha$
(FP2) $\lambda : A \vee B \leftrightarrow \lambda : A \vee \lambda : B$, where $\lambda > \alpha$
(FP3) $\lambda : A \to B \leftrightarrow \lambda : A \to \lambda : B$, where $\lambda > \alpha$
(FP4) $\lambda : A \wedge B \leftrightarrow \lambda : A \vee \lambda : B$, where $\lambda < \alpha$
(FP4) $\lambda : A \vee B \leftrightarrow \lambda : A \wedge \lambda : B$, where $\lambda < \alpha$
(FP5) $\lambda : A \to B \leftrightarrow (1 - \lambda) : A \wedge \lambda : B$, where $\lambda < \alpha$
(FP6) αp

Fuzzy annotated logics $FP\tau$ can be seen as a fragment of annotated logics $P\tau$ in that the former can be translated into the latter. We call a formula built from boolean connection of fuzzy atomic formulas a *basic formula*. Formulas which are not basic are called *non-basic formulas*. Then, we have:

Theorem 5.1 *Any non-basic formulas in* $FP\tau$ *can be transformed into basic formulas in* $FP\tau$.

Proof Immediate from the axioms (FP1)–(FP6).

Next, we define the translation t of basic formulas in $FP\tau$ to formulas in $FP\tau$ as follows:

$t(\lambda : p) = p_\lambda$ if $\lambda > \alpha$,
$t(\lambda : p) = p_\perp$ if $\lambda = \alpha$,
$t(\lambda : p) = \neg p_\lambda$ if $\lambda < \alpha$,
$t(A \wedge B) = t(A) \wedge t(B)$,
$t(A \vee B) = t(A) \vee t(B)$,
$t(A \rightarrow B) = t(A) \rightarrow t(B)$.

The translation t can reinterpret the annotation in $FP\tau$ as the negation $\neg$ in $P\tau$. Then, we can reach the following theorem:

Theorem 5.2 *Any formulas A in $FP\tau$ can be translated as $t(A)$ into $P\tau$.*

Proof By using Theorem 5.1, we can prove the statement by induction on the length of A.

As a consequence of Theorem 5.2 and the completeness theorem of $P\tau$ (Theorem 2.17), the completeness of $FP\tau$ follows. Fuzzy annotated logics are suited to deal with approximated reasoning as reasoning in annotated logics and unlike many systems of fuzzy logic they have advantages with some logical properties.

As said before, our systems are similar to the fuzzy operator logics of Weigert et al. [112, 158] who presented the theory and resolution procedure of the logics. Fuzzy operator logic also allows the fuzzy operator of a formula, in which the fuzzy operator is regarded as a kind of negation, and we can point out some similarities of $FP\tau$ and fuzzy operator logics. However, the semantics of the fuzzy operator logic needs a kind of fuzzy product $\otimes$ defined as:

$$\phi \otimes \delta = (2\phi - 1) \cdot \delta - \phi + 1$$

In addition, fuzzy operator logic does not have implication. Weigert, Tsai and Liu's works did not provide the axiomatization of fuzzy operator logics. But we believe that fuzzy operator logic can be formalized as a version of fuzzy annotated logics.

5.2 Possibilistic Annotated Logics

Possibilistic annotated logics were developed by Akama and Abe [26] to extend annotated logic with the degree of inconsistency motivated by *possibilistic logic* developed by Dubois and Prade. As discussed, paraconsistent logics including annotated logics can tolerate inconsistency.

However, we must recognize that there two kinds of inconsistency in a logical system, i.e. *local inconsistency* and *global inconsistency*. Local inconsistency is a negation-inconsistency of the form $A \wedge \neg A$, where $\neg$ denotes negation and $\wedge$ conjunction, respectively. Global inconsistency means *triviality* in the sense that every formula is derivable. In paraconsistent logics, local inconsistency does not trigger global inconsistency.

However, in most logical systems like classical and intuitionistic logic, local inconsistency implies global inconsistency. The feature is not desirable for practical applications because real systems often face local inconsistency. In this regard, paraconsistent logics are useful as a foundation for real systems capable of handling inconsistent information.

Now, a problem arises. Indeed paraconsistent logics can tolerate local inconsistency but they give no idea of how to handle (or interpret) it. The important thing in applications is to tell what a system should do after finding local inconstancy. Thus, the use of current paraconsistent logics is not enough.

We need an additional mechanism for tolerating local inconsistency. Possibilistic annotated logics $PP\tau$ are considered by fusing annotated logics $P\tau$ and possibilistic logic. We believe that the possibilistic extension of annotated logics is interesting because several contradictions can be distinguished by a necessity degree.

Before setting up possibilistic annotated logics $PP\tau$, we survey possibilistic logic; see Dubois et al. [75] for details. Possibilistic logic is a logic inspired by *possibility theory* of Zadeh [160]. It can deal with possibility and necessity measure of a formula in some logical system. Here, necessity measure can express certainty degree in the sense of measuring the certainty of a formula to be true. In addition, possibilistic logic can describe *partial* inconsistencies. And it seems to be closely related to the ideas in paraconsistent logics.

We here provide a formal sketch of possibilistic logic *PL*. Let *L* be the language of classical propositional logic *CL*. Let Ω be the set of interpretations of *L*. A *possibility distribution* on Ω is a function π from Ω to [0, 1]. π is called *normalized* iff $\exists\omega \in \Omega$ such that $\pi(\omega) = 1$. The quantity $SN(\pi) = 1 - sup\{\pi(\omega) \mid \omega \in \Omega\}$ is called the *subnormalized degree* of π, and is equal to 0 iff π is normalized.

A possibility distribution π on Ω provides two types of functions on the set L' of formulas of *L* to [0, 1], called *possibility measure* Π and *necessity measure* N, defined as follows:

$$\begin{aligned}\Pi : L' \to [0, 1], (\forall p \in L')\Pi(p) &= Sup\{\pi(\omega) \mid \omega \models p\}),\\ N : L' \to [0, 1], (\forall p \in L')N(p) &= Inf\{1 - \pi(\omega) \mid \omega \models \neg p\}\\ &= 1 - \Pi(\neg p).\end{aligned}$$

Here, we note that possibilistic logic should be distinguished from *fuzzy logic* in that fuzzy logic is a logic of vagueness.

In *PL*, we use a *necessity-valued formula* of the form $(p\ \alpha)$, where p is a propositional formula and α is a valuation of [0, 1]. The necessity-valued formula $(p\ \alpha)$ states that $N(p) \geq \alpha$, namely the satisfaction of p is at least α-necessity. Thus, $(p\ 1)$ expresses that $N(p) \geq \alpha$, namely p must be absolutely satisfied. And, $(p\ 0)$ expresses that p is not useful, namely we know nothing about p.

We turn to the semantics of *PL*. The standard formula can be interpreted in the usual way. For necessity-valued formulas, the satisfaction relation $\models$ is defined as follows:

$\pi \models (p\ \alpha)$ iff $N(p) \geq \alpha$

where π is a possibility distribution on Ω and N is the necessity measure induced by π. Let $\mathscr{F}$ be the set of necessity-valued formulas $\{(p_1,\ \alpha_1), \ldots, (p_n\ \alpha_n)\}$. Then, we can define the following:

$\pi \models \mathscr{F}$ iff $\forall i \in \{1, \ldots, n\}, \pi \models (p_i\ \alpha_i)$

The notion of logical consequence is defined as:

$\mathscr{F} \models (p\ \alpha)$ iff $\forall \pi (\pi \models \mathscr{F}$ imply $\pi \models (p\ \alpha))$

A Hilbert-style axiomatization of *PL* is presented by the following postulates:

Postulates of ***PL***

(PL1) $(A \rightarrow (B \rightarrow A)\ 1)$
(PL2) $((A \rightarrow (B \rightarrow C) \rightarrow ((A \rightarrow B) \rightarrow (A \rightarrow C))\ 1)$
(PL3) $((\neg A \rightarrow \neg B) \rightarrow ((\neg A \rightarrow B) \rightarrow A)\ 1)$
(GMP) $(A\ \alpha), (A \rightarrow B\ \beta) \vdash (B\ min(\alpha, \beta))$
(S) $(p\ \alpha) \vdash (p\ \beta)$, where $\beta \leq \alpha$

Here, (PL1)–(PL3) denote axioms, and (GMP) and (S) are rules of inferences. $\vdash$ is the provability relation in *PL*. For the axioms, we can use any axiomatics of the classical propositional logic *PC*. (GMP) is the *graded modus ponens*, which is a possibilistic generalization of *modus ponens* in *PC*. (S) is a rule for the necessity measure, which is similar to the necessitation in modal logic.

We can prove the completeness theorem of *PL*; see Dubois et al. [75]. One of the distinguished features of *PL* is to measure the consistency of formulas. The models of a set of necessity-valued formulas $\mathscr{F}$ are possibility distributions on the set Ω of all interpretations for L.

Thus, we can define the *consistency degree* of $\mathscr{F}$, denoted $Cons(\mathscr{F})$, as follows:

$Cons(\mathscr{F}) = Sup_{\pi \models \mathscr{F}} Sup_{\omega \in \Omega} \pi(\omega)$

The dual of $Cons(\mathscr{F})$ is the *inconsistency degree*, denoted $Incons(F)$, defined as follows:

$Incons(\mathscr{F}) = 1 - Sup_{\pi \models \mathscr{F}} Sup_{\omega \in \Omega} \pi(\omega)$

We here observe that *PL* has the paraconsistent feature by extending classical logic in the following sense. Let $F = \{p_i \mid i = 1, \ldots, n\}$ be a set of classical formulas and let $\mathscr{F} = \{(p_i\ 1) \mid i = 1, \ldots, n\}$. Then, we have that if $\mathscr{F}$ is consistent then $Incons(\mathscr{F}) = 0$ and that if $\mathscr{F}$ is inconsistent then $Incons(\mathscr{F}) = 1$.

Moreover, we can grade the inconsistency of formulas. We say that $\mathscr{F}$ is *completely consistent* if $Incons(\mathscr{F}) = 0$. $\mathscr{F}$ is called *completely inconsistent* if $Incons(\mathscr{F}) = 1$. $\mathscr{F}$ is *partially inconsistent* if $0 < Incons(\mathscr{F}) < 1$.

Thus, we can measure the global inconsistency of the formula $\mathscr{F}$ using $Incons(\mathscr{F})$. We describe several results about these notions.

Theorem 5.3 *Let N be the necessity distribution induced by π and $\perp$ be the contradiction. Then, we have.*

$Incons(\mathscr{F}) = Inf\{N(\perp) \mid \pi \models \mathscr{F}\}$.

Proof This property can be proved as follows:

$N(\perp) = Inf\{1 - \pi(\omega) \mid \omega \not\models \perp\} = Inf_{\omega \in \Omega}(1 - \pi(\omega)) = SN(\pi)$

Theorem 5.3 says that the inconsistency degree is the minimal necessity degree of the contradiction $\perp$ for all possibility distributions satisfying $\mathscr{F}$.

Theorem 5.4 *Let* $\mathscr{F} = \{(p_1\, \alpha_1), \ldots, (p_n\, \alpha_n)\}$ *be a set of necessity-valued formulas, and let us define the possibility distribution* $\pi_{\mathscr{F}}$ *as*

$$\pi_{\mathscr{F}} = Inf\{1 - \alpha_i \mid \omega \models \neg\alpha_i, i = 1, \ldots, n\}.$$

Then, π *satisfies* $\mathscr{F}$ *iff* $\pi \leq \pi_{\mathscr{F}}$ *for any possibility distribution* π *on* Ω.

Proof Assume that π satisfies $\mathscr{F}$. Then, it is equivalent to the following:

$$(\forall i = 1, \ldots, n)\pi \models (p_i\, \alpha_i).$$

If we write N for the necessity measure induced by π then it is equivalent to:

$$\begin{aligned}
&(\forall i = 1, \ldots, n)N(p_i) \geq \alpha_i \\
&\Leftrightarrow\ (\forall i = 1, \ldots, n)Inf\{1 - \pi(\omega) \mid \omega \models \neg p_i\} \geq \alpha_i \\
&\Leftrightarrow\ (\forall i = 1, \ldots, n)(\forall \omega \models\models \neg p_i)\pi(\omega) \leq 1 - \alpha_1 \\
&\Leftrightarrow\ \forall \omega \in \Omega, \pi(\omega) \leq Inf\{1 - \alpha_1 \mid \omega \models \neg p_i, i = 1, \ldots, n\} \\
&\Leftrightarrow\ \forall \omega \in \Omega, \pi(\omega) \leq \pi_{\mathscr{F}}(\omega).
\end{aligned}$$

From Theorem 5.4, we have:

Theorem 5.5 $\mathscr{F} \models (p\, \alpha)$ *iff* $\pi_{\mathscr{F}} \models (p\, \alpha)$.

Proof Let N and N' be possibility distributions induced by π and π', respectively. Then, by the definition of necessity measure, we have that $\pi \leq \pi'$ iff $N \geq N'$. That is,

$$\begin{aligned}
&\mathscr{F} \models (p\, \alpha) \text{ iff } \forall \pi, \pi \text{ entails } N(p) \geq \alpha \\
&\Leftrightarrow\ \forall \pi, \pi \leq \pi_{\mathscr{F}} \text{ entails } N(p) \geq \alpha \\
&\Leftrightarrow\ \forall \pi, N \geq N_{\mathscr{F}} \text{ entails } N(p) \geq \alpha \\
&\Leftrightarrow\ N_{\mathscr{F}(p)} \geq \alpha \\
&\Leftrightarrow\ \pi_{\mathscr{F}} \text{ satisfies } \models (p\, \alpha).
\end{aligned}$$

As a consequence, we have the relation of inconsistency degree and subnormalized degree.

Theorem 5.6 $Incons(\mathscr{F}) = 1 - Sup_{\omega \in \Omega}\pi_{\mathscr{F}}(\omega) = SN(\pi_{\mathscr{F}})$.

By Theorem 5.6, we can evaluate the inconsistency degree of $\mathscr{F}$ as the inconsistency degree of the subnormalization of the possibility distribution $\pi_{\mathscr{F}}$.

We turn to an exposition of possibilistic annotated logics $PP\tau$. Possibilistic annotated logics are formulated by amalgamating annotated and possibilistic logics, in which both local and global inconsistency can be expressed. The idea is inspired by the work in Besnard and Lang [51]. They claimed that non-classical logics can be strengthened by possibilistic logic.

As a case study, they treated a possibilistic extension of da Costa's [61] C_1 and developed a notion of *graded paraconsistency*. The choice of C_1 seemed to be made by the historical reason that C_1 is one of the well known paraconsistent logics.

However, they claimed that their approach could be applied to *any* paraconsistent logics.

In this regard, to develop a possibilistic annotated logic $PP\tau$ is very interesting. This is because annotated logics are computational logic. We also note that annotated logics employ the device of annotation and share similar ideas with possibilistic logics. Below, we work out the new system and discuss the advantages over other approaches.

The language of $PP\tau$ is that of $P\tau$ with the necessity-valued formulas of the form: $(\alpha\ p)$, where $\alpha \in N$ is a valuation of [0, 1] and p is a formulas in $P\tau$. It is therefore possible to give a possibilistic interpretation of the formulas of $P\tau$. Here, N is a $P\tau$-necessity function from L of the language of $P\tau$ to [0, 1] satisfying the following three conditions:

$(Taut)$ if $\vdash_{P\tau} A$ then $N(A) = 1$
(Eq_N) if $\vdash_{P\tau} A \leftrightarrow B$ then $N(A) = N(B)$
$(Conj)$ $N(A \wedge B) = min(N(A), N(B))$

Consequently, we can also provide a $P\tau$-possibility function Π satisfying the following:

$(Contr)$ if $\vdash_{P\tau} \neg_* A$ then $\Pi(A) = 0$
(Eq_Π) if $\vdash_{P\tau} A \leftrightarrow B$ then $\Pi(A) = \Pi(B)$
$(Disj)$ $\Pi(A \vee B) = max(\Pi(A), \Pi(B))$

Observe here that $(Contr)$ holds only for strong negation $\neg_*$. It is easily verified that $P\tau$-necessity function satisfies the following theorem.

Theorem 5.7 *Let N be a $P\tau$-necessity function. Then, N satisfies:*

$(N1)$ *if* $\vdash_{P\tau} A \rightarrow B$ *then* $N(B) \geq N(A)$
$(N2)$ $N(B) \geq min(N(A), N(A \rightarrow B))$
$(N3)$ *if* $A_1, \ldots, A_n \vdash_{P\tau} B$ *then* $N(B) \geq min(N(A_1), \ldots, N(A_n))$

Proof For $(N1)$, $\vdash_{P\tau} A \leftrightarrow (A \wedge B)$ if $\vdash_{P\tau} A \rightarrow B$. From (Eq_N), $N(A) = N(A \wedge B)$ holds. Next, by $(Conj)$, we have $N(A \wedge B) = min(N(A), N(B))$. Thus, $N(B) \geq N(A)$.

For $(N2)$, the following two formulas are provable from (A5) and (A6) from a Hilbert system of $P\tau$.

(1) $\vdash_{P\tau} A \wedge (A \rightarrow B) \rightarrow A$
(2) $\vdash_{P\tau} A \wedge (A \rightarrow B) \rightarrow (A \rightarrow B)$

By (A2), we have:

(3) $\vdash_{P\tau} (A \wedge (A \rightarrow B) \rightarrow (A \rightarrow B))$
$\rightarrow ((A \wedge (A \rightarrow B) \rightarrow A) \rightarrow (A \wedge (A \rightarrow B) \rightarrow B))$

From (2), (3), by using (A4)

(4) $\vdash_{P\tau} (A \wedge (A \rightarrow B) \rightarrow A) \rightarrow (A \wedge (A \rightarrow B) \rightarrow B)$

is obtained. From (1) and (4), we get:

(5) $\vdash_{P\tau} A \wedge (A \rightarrow B) \rightarrow B$

By (A4). From (5), $N(B) \geq N(A \wedge (A \rightarrow B))$ holds by $(N1)$. Next, by applying $(Conj)$ to it, we can derive $N(B) \geq min(N(A), N(A \rightarrow B))$.

For (N3), from $A_1, \ldots, A_n \vdash_{P\tau} B$, we have $\vdash_{P\tau} A_1 \wedge \cdots \wedge A_n \rightarrow B$ by the deduction theorem. By $(N1), N(B) \geq N(A_1 \wedge \cdots \wedge A_n)$ holds. Next, $(Conj)$ enables us to deduce $N(A_1 \wedge \cdots \wedge A_n) = min(N(A_1), \ldots, N(A_n))$. Finally, by $(N1)$, we obtain $N(B) \geq min(N(A_1), \ldots, N(A_n))$.

Next, we give a semantics for $PP\tau$. The interpretation $I : \mathbf{P} \rightarrow \mid \tau \mid$ and the associated valuation $V_I : \mathbf{F} \rightarrow \mathbf{2}$ are defined as in $P\tau$. In addition, we must specify the valuation $V_N : N(\mathbf{F}) \rightarrow \mathbf{2}$ for necessity-valued formulas as follows:

$V_N((A\ \alpha)) = 1$ if $N(A) \geq \alpha$
$V_N((A\ \alpha)) = 0$ otherwise

The axiomatization of $PP\tau$ is obtainable from the axiomatization of $P\tau$ by replacing each axiom ϕ of $P\tau$ by $(\phi\ 1)$ except $(\rightarrow_4)$, i.e. *modus ponens*. It is modified as a possibilistic version (PA4) and the additional rule (PA18) is added. Then, the postulates of $PP\tau$ are presented as follows:

Postulates of $PP\tau$

(PA1) $(A \rightarrow (B \rightarrow A)\ 1)$
(PA2) $((A \rightarrow (B \rightarrow C)) \rightarrow ((A \rightarrow B) \rightarrow (A \rightarrow C))\ 1)$
(PA3) $((A \rightarrow B) \rightarrow A) \rightarrow A\ 1)$
(PA4) $(A\ \alpha), (A \rightarrow B\ \beta)/(B\ min(\alpha, \beta))$
(PA5) $(A \wedge B \rightarrow A\ 1)$
(PA6) $(A \wedge B \rightarrow B\ 1)$
(PA7) $(A \rightarrow (B \rightarrow (A \wedge B))\ 1)$
(PA8) $(A \rightarrow A \vee B\ 1)$
(PA9) $(B \rightarrow A \vee B\ 1)$
(PA10) $((A \rightarrow C) \rightarrow ((B \rightarrow C) \rightarrow ((A \vee B) \rightarrow C))\ 1)$
(PA11) $((F \rightarrow G) \rightarrow ((F \rightarrow \neg G) \rightarrow \neg F)\ 1)$
(PA12) $(F \rightarrow (\neg F \rightarrow A)\ 1)$
(PA13) $(F \vee \neg F\ 1)$
(PA14) $(p_\perp\ 1)$
(PA15) $(\neg^k p_\lambda \rightarrow \neg^{k-1} p_{\sim\lambda}\ 1)$, where $k \geq 1$
(PA16) $(p_\lambda \rightarrow p_\mu\ 1)$, where $\lambda \geq \mu$
(PA17) $(p_{\lambda_1} \wedge p_{\lambda_2} \wedge \cdots \wedge p_{\lambda_m} \rightarrow p_\lambda\ 1)$, where $\lambda = \bigvee_{i=1}^{m} \lambda_i$
(PA18) $(A\ \alpha)/(A\ \beta)$, where $\beta \leq \alpha$

Here, (PA4) is (GMP) and (PA18) is (S) in the sense of possibilistic logic.

$PP\tau$ extends $P\tau$ in several respects. Since $P\tau$ is a paraconsistent logic, it can tolerate local inconsistency at the atomic level, namely hyper-literals. But, $PP\tau$ can also express global inconsistency at all levels of formulas by means of the necessity measure. Consequently, we can grade the level of inconsistency $(A \wedge \neg_* A)$.

Now, we prove the completeness theorem of $PP\tau$. The proof is regarded as a possibilistic extension of that of $P\tau$. Thus, the result is derived from the completeness of $P\tau$ and that of PL. However, we here provide the proof hence making this paper self-contained. First, we note that the *deduction theorem* holds in $PP\tau$.

Theorem 5.8 $\mathscr{F} \cup \{(A\ 1)\} \models (B\ \alpha) \Leftrightarrow \mathscr{F} \models (A \rightarrow B\ \alpha)$, *where* $\models$ *is shorthand for* $\models_{PP\tau}$.

Proof For ($\Rightarrow$), Suppose $\mathscr{F} \cup \{(A\ 1)\} \models_{PP\tau} (A\ \alpha)$ holds. By Theorem 5.5, it implies:

$$N_{\mathscr{F} \cup \{(A\ 1)\}}(B) \geq \alpha.$$

This is transformed as follows:

$$\Rightarrow Inf\{1 - \pi_{\mathscr{F} \cup (A\ 1)}\}(\omega), \omega \models \neg_* B\} \geq \alpha$$
$$\Rightarrow \forall \omega \models A \wedge \neg_* B, \pi_{\mathscr{F}}(\omega) \leq 1 - \alpha$$

Here, the second transformation is justified, because $\pi_{\mathscr{F} \cup \{(A\ 1)\}}(\omega) = \pi_{\mathscr{F}}(\omega)$ if $\omega \models B$. Next, we have that:

$$N_{\mathscr{F}}(A \rightarrow B) \geq \alpha$$

Thus, from Theorem 5.5, it follows that:

$$\mathscr{F} \models (A \rightarrow B\ \alpha)$$

For ($\Leftarrow$), assume that $\mathscr{F} \models (A \rightarrow B\ \alpha)$. By the definition of the necessity-valued formula, we have:

$$\forall \pi \models \mathscr{F}, N(A \rightarrow B) \geq \alpha.$$

Here, since $N(B) \geq min(N(A), N(A \rightarrow B))$ it follows that:

$$\forall \pi \models \mathscr{F}, N(A) = 1 \text{ implies } N(B) \geq \alpha.$$

This is further transformed into the following:

$$\Rightarrow \forall \pi \models \mathscr{F} \cup \{(A\ 1)\}, N(B) \geq \alpha$$
$$\Rightarrow \mathscr{F} \cup \{(A\ 1)\} \models (B\ \alpha)$$

as desired.

Now, we turn to the completeness of $PP\tau$. Its proof uses the completeness of $P\tau$. Armed with the completeness of $P\tau$ (Theorem 2.12), we can obtain the completeness theorem for $PP\tau$.

Theorem 5.9 (Completeness) *Let A be a formula in* $P\tau$ *and* $\mathscr{F}$ *be a set of necessity-valued formulas. Then, we have:*

$$\mathscr{F} \vdash_{PP\tau} (A\ \alpha) \textit{ iff } \mathscr{F} \models_{PP\tau} (A\ \alpha)$$

Proof The soundness is trivial. For completeness, assume that $\mathscr{F} \models_{PP\tau} (A\,\alpha)$. Then, it is equivalent to the following:

$$\mathscr{F}_\alpha \models A$$

where $\mathscr{F}_\alpha = \{B_i \mid (B_i\,\alpha_i) \in \mathscr{F}, \alpha_i \geq \alpha\}$. This equivalence is derivable from the definition of necessity-valued formula and that of entailment. The completeness of $P\tau$ ensures that $\mathscr{F}_\alpha \vdash_{P\tau} A$ for any non-necessity-valued formula A. This implies that $\mathscr{F}_\alpha \vdash_{PP\tau} (A\,\beta)$, where $\beta \geq \alpha$. Then, we can conclude that $\mathscr{F} \vdash_{PP\tau} (A\,\alpha)$ by means of (PA18).

In this way, we can exploit the completeness of $P\tau$. However, the result seems uniform because the argument depends on the completeness of the base system $P\tau$. This suggests that the possibilistic extension considered here is universal in that the construction can be achieved for any paraconsistent (also non-classical) logics. But, as far as we aware, the completeness proof in this line is firstly detailed.

$PP\tau$ is another possibilistic extension of paraconsistent logics as Besnard and Lang [51] considered it for da Costa's C_1. The proposed logics are more powerful and computational than the corresponding system based on C_1 in the sense that we dispense with the sophisticated treatment of paraconsistent negation and that they have a computational adequacy.

5.3 Inductive Annotated Logics

Inductive annotated logics were proposed by da Costa and Krause [67] as an extension of annotated logics capable of various forms of reasoning such as non-monotonic and defeasible reasoning. They criticized logical systems which only concern non-doxastic states of inputs and outputs about the states of the data because there are no references to degrees of belief or confidence, and developed inductive annotated logics $I\tau$ adding the postulates for degrees of confidence to annotated logics $P\tau$.

Below we review inductive annotated logics following da Costa and Krause [67]. Their starting point is to take into account both degrees of *vagueness* and degrees of *confidence*. A proposition may have a certain 'degree of vagueness'. For example, 'Peter is smart' is a vague proposition. Such a proposition could be expressed by a certain logic for vagueness like fuzzy logic.

However, the formalization in the logic for vagueness does not appear enough since for common-sense reasoning we need a degree of confidence about the proposition. This means that the desired representation of 'Peter is smart' is to formalize 'we believe that Peter is smart with some degree of confidence'.

Annotated logics can serve as a logic for vagueness. But, to implement da Costa and Krause's idea for representing propositions, annotated logics should be extended so that the degrees of confidence are attached to propositions. The resulting systems are inductive annotated logics $I\tau$.

Their proposed logics are general enough to deal with various logics for uncertainty in the literature. The fuzzy annotated logics and possibilistic annotated logics presented above could be interpreted as versions of inductive annotated logics.

Here is a formal definition of inductive annotated logics whose language is denoted by L. Inductive annotated logics are based on two kinds of (complete) lattices, i.e., a lattice of vagueness denoted $\tau = \langle | \tau |, \leq \rangle$ and a lattice of confidence denoted by $\sigma = \langle | \sigma |, \leq \rangle$. We write $\bot$ and $\top$ the bottom and top element of τ and $\bot'$ and $\top'$ those of σ. We use $\vee$ and $\wedge$ for the least upper bound and the greatest lower bound both for τ and σ.

We assume that $\mu_1, \mu_2, \ldots \in \tau$ and that $\lambda_1, \lambda_2, \ldots \in \sigma$. Note that da Costa and Krause used a particular finite linearly ordered set $\tau = \{\mu_1, \ldots, \mu_4\}$ where $\mu_1 (= \bot) \leq \cdots \leq \mu_4 (= \top)$ but that the choice is for their exposition.

Definition 5.4 (*Symbols*) The symbols of $P\tau$ are defined as follows:

1. Propositional symbols: $p, q, \ldots$ (possibly with subscript)
2. Annotated vague constants: $\mu_1, \mu_2, \ldots \in | \tau |$
3. Annotated confidence constants: $\lambda_1, \lambda_2, \ldots \in | \sigma |$
4. Logical connectives: $\wedge$ (conjunction), $\vee$ (disjunction), $\rightarrow$ (implication), and $\neg$ (negation)
5. Parentheses: (and)

Definition 5.5 (*Formulas*) Formulas are defined as follows:

1. If p is a propositional symbol and $\mu \in | \tau |$ is an annotated constant, then p_μ is a formula (an annotated atom).
2. If F is a formula, then $\neg F$ is a formula.
3. If F and G are formulas, then $F \wedge G$, $F \vee G$, $F \rightarrow G$ are formulas.
4. If p is a propositional symbol and $\mu \in | \tau |$ is an annotated vague constant, then a formula of the form $\neg^k p_\mu$ $(k \geq 0)$ is a hyper-literal. A formula which is not a hyper-literal is a complex formula.
5. If F is a formula and $\lambda \in | \sigma |$ is an annotated confidence constant, then $F : \lambda$ is a formula.

Observe here that annotated vague constants can only be attached to a propositional symbol but that annotated confidence constants can be attached to any formula. Strong negation $\neg_*$ can be introduced as in $P\tau$.

Next, we describe a semantics for $I\tau$, consisting of a semantics for $P\tau$ and that for degrees of confidence. Although the former was presented in Chap. 2, we repeat it for our purpose here. We write $\mathbf{P}$ for the set of propositional variables and $\mathbf{F}$ the set of formulas, and $\mathbf{2} = \{0, 1\}$ the set of truth-values, respectively.

The image of the proposition p by the mapping h is denoted by p_μ, where $\mu \in | \tau |$. The informal meaning of a formula p_μ is that p is true with degree of vagueness μ. For each interpretation h, we define a valuation $v_h : \mathbf{F} \rightarrow \mathbf{2}$.

Definition 5.6 (*Semantics*) Let h be a mapping, v_h be a valuation, $p \in \mathbf{P}$, and A and B be formulas. Then,

1. $v_h(p_\mu) = 1 \Leftrightarrow \mu \leq h(p)$.
2. $v_h(\neg^k p_\mu) = v_h(\neg^{k-1} p_{\sim\mu})$, where $k \neq 0$.
3. $v_h(A \wedge B) = 1 \Leftrightarrow v_h(A) = v_h(B) = 1$.
4. $v_h(A \vee B) = 1 \Leftrightarrow v_h(A) = 1$ or $v_h(B) = 1$.
5. $v_h(A \rightarrow B) = 1 \Leftrightarrow v_h(A) = 0$ or $v_h(B) = 1$.
6. If F is a complex formula, then $v_h(\neg F) = 1 \Leftrightarrow v_h(F) = 0$.

We say that v_h *satisfies* A if $v_h(A) = 1$ and that it does not satisfy A if $v_h(A) = 0$. If Γ is a set of formulas, then we say that a formula A is a *semantic consequence* of Γ, written $\Gamma \models A$, iff for every valuation v_h such that $v_h(B) = 1$ for each $B \in \Gamma$, then $v_h(A) = 1$. A formula A is *valid*, written $\models A$, iff $\emptyset \models A$. A valuation v_h is a *model* for a set of formulas Γ iff $v_h(B) = 1$ for every $B \in \Gamma$. v_h is a model of A iff $v_h(A) = 1$.

Degrees of confidence of propositions (including vague propositions) can be formalized by a *theory of confidence*. We can attribute degrees of confidences to propositions by a confidence function.

Definition 5.7 (*Confidence function*) Let A and B be formulas and $|\sigma|$ be a lattice of confidence. A *confident function* is a mapping $\mathbf{C} : \mathbf{F} \rightarrow |\sigma|$ satisfying the following[1]:

1. $\mathbf{C}(A \wedge \neg_* A) = \bot'$
2. $\mathbf{C}(A \vee \neg_* A) = \top'$
3. $\mathbf{C}\left(\bigvee_{i \in I} A_i\right) \geq \bigvee_{i \in I} \mathbf{C}(A_i)$, for I finite.
4. $\mathbf{C}\left(\bigwedge_{i \in I} A_i\right) \leq \bigwedge_{i \in I} \mathbf{C}(A_i)$, for I finite.
5. if $\models A \leftrightarrow B$, then $\mathbf{C}(A) = \mathbf{C}(B)$.

It is noted that the definition of the confidence function can be modified in various ways to implement concepts like probability and possibility in some theories.

We turn to a proof theory for $I\tau$. For this, we use the convention $p_\mu : \lambda$ for $\mathbf{C}(p_\mu) = \lambda$. $I\tau$ needs postulates for $P\tau$ and the additional rule called the *warning rule* (*WR*).

Postulates of $I\tau$

($P\tau$) Postulates for $P\tau$

$$(WR)\ \frac{p_{\mu_i} : \lambda_i,\ p_{\mu_j} : \lambda_j}{p_{\mu_i \wedge \mu_j} : \lambda_i \vee \lambda_j}\ \Gamma$$

---

[1] We use $\bigvee$ ($\bigwedge$) both for the logical connective and the lattice-theoretical operation.

Here, Γ stands for the set of side conditions which are the underlying base for the application of the rule, conditions that are accepted as true.

The warning rule can handle various rules like induction rules and non-monotonic rules with degrees of vagueness and confidence. Intuition behind the warning rule reveals that a possible expert system elaborated for dealing with vagueness and with degrees of confidence.

Although da Costa and Crause's exposition is somewhat informal, inductive annotated logics $I\tau$ can serve as a general framework for dealing with a wider range of uncertain reasoning. It is also noticed that we can automate the inferences in these logics by means of the proof method for annotated logics. However, more work should be done to evaluate their usefulness.

5.4 Structural Annotated Logics

Structural annotated logics were studied by Lewin et al. [109, 110] to improve shortcomings of algebraic aspects of the original version of annotated logics, i.e., $P\tau$. Lewin, Mikenberg and Schwarze proposed structural annotated logics $SP\tau$ which are equivalent to $P\tau$ in [110] and later they developed structural annotated logics SAL_τ which are axiomatic extensions of $SP\tau$.

Theoretically, these systems have two important features. One is that they are *structural* in the sense that their consequence relation is closed under substitution. The other is that we can find their algebraic counterparts. Annotated logics $P\tau$ lack these features in a strict sense.

First, we give the details of $SP\tau$ following Lewin et al. [109]. As mentioned above, their annotated logics were motivated by overcoming two difficulties of $P\tau$. It can be shown that $P\tau$ and $SP\tau$ are equivalent in the sense that both systems prove the "same" formulas in the corresponding language.

Definition 5.8 (*Symbols*) The symbols of $SP\tau$ are defined as follows:

1. Propositional symbols: $p, q, \ldots$ (possibly with subscript)
2. Annotated constants: $\mu, \lambda, \ldots \in |\tau|$
3. Logical connectives: $\wedge$ (conjunction), $\vee$ (disjunction), $\rightarrow$ (implication), $\neg$ (negation), $^\circ$, and f_λ for each $\lambda \in |\tau|$.
4. Parentheses: (and)

The language of $SP\tau$ extends that of $P\tau$ with $^\circ$ and f_λ.

Definition 5.9 (*Formulas*)

1. Propositional symbols $p, q, r, \ldots$ are formulas.
2. If A is a formula, then A° is a formula.
3. If A is a formula and λ is an annotation constant, then $f_\lambda A$ is a formula.

4. If A and B are formulas, then $\neg A, A \wedge B, A \vee B$ and $A \rightarrow B$ are formulas.
5. If p is a propositional symbol and $\lambda_i \in |\tau|$ $(0 \leq i \leq n, k_i \in \omega$ (the set of natural numbers)), then $\neg^{k_n} f_{\lambda_n} \neg^{k_{n-1}} f_{\lambda_{n-1}} \ldots \neg^{k_1} f_{\lambda_1} \neg^{k_0} p$ is a formula called a hyper-literal. A formula that is not a hyper-literal is called a complex formula.

A formula A° reads "A is well-behaved". As in $P\tau$, the function $\sim$ is used for hyper-literal. If τ is a finite lattice, then for any $\lambda \in |\tau|$, the set $\{\sim^m \lambda \mid m \in \omega\}$ is a finite set. For any λ, there exists a least integer t_λ such that for some p:

$$\sim^{t_\lambda} \lambda = \sim^{t_\lambda + p} \lambda.$$

Of course there will also exist the least such an integer p. We call it p_λ. If $r \geq t_\lambda$ and s is a multiple of p_λ, then we have:

$$\sim^r \lambda = \sim^{r+s} \lambda.$$

Then, we define:

$$T = max\{t_\lambda \mid \lambda \in |\tau|\}$$
$$P = m.c.m.(\{2\} \cup \{p_\lambda : \lambda \in |\tau|\})$$

where $m.c.m.$ denotes the minimum common multiple, then we have, for any $\lambda \in |\tau|$,

$$\sim^T \lambda = \sim^{T+P} \lambda.$$

Now, we are ready to describe a semantics for $SP\tau$ which is similar to that for $P\tau$. Let $\mathscr{P}$ be the set of propositional symbols, $\mathscr{F}$ the set of formulas and $\mathbf{2} = \{0, 1\}$.

Definition 5.10 (*Valuation*) An *interpretation* is a function $I : \mathscr{P} \rightarrow |\tau|$. For each interpretation I, we define a *valuation* $v_I : \mathscr{F} \rightarrow \mathbf{2}$ as follow:

1. $v_I(p) = 1 \Leftrightarrow I(p) \geq \top$.
2. $v_I(\neg^k p) = 1 \Leftrightarrow I(p) \geq \sim^k \top$.
3. $v_I(A^\circ) = 1 \Leftrightarrow A$ is not a hyper-literal.
4. If $A = \neg^{k_1} f_{\lambda_1} \neg^{k_2} f_{\lambda_2} \ldots \neg^{k_n} f_{\lambda_n} \neg^{k_{n+1}} p$, then
   $v_I(f_\lambda A) = 1 \Leftrightarrow I(p) \geq \lambda$.
5. If A is a complex formula, then
   $v_I(f_\lambda) = v_I(A)$.
6. If A is a hyper-literal, then
   $v_I(\neg^k f_\lambda A) = v_I(\neg^{k-1} f_{\sim\lambda} A)$.
7. If A is a complex formula, then
   $v_I(\neg A) = 1 - v_I(A)$
8. For binary connectives, v_I is defined as in $P\tau$.

The notions of truth, validity and semantic consequence are defined as in $P\tau$.

Next, we show an axiomatization of $SP\tau$. Note that the systems are very different if the lattice τ is finite or infinite.

Postulates for $SP\tau$

1. The axioms for $\wedge$, $\vee$ and $\to$ are the same as those for $P\tau$.
2. Axioms for negation

   $(\neg_{1S})$ $(A^\circ \wedge B^\circ) \to ((A \to B) \to ((A \to \neg B) \to \neg A))$
   $(\neg_{2S})$ $A^\circ \to (A \to (\neg A \to B))$
   $(\neg_{3S})$ $A^\circ \to (A \vee \neg A)$
   $(\neg_{4S})$ $\neg^{T+P} A = \neg^T A$

   where T and P are the defined above. They depend on the lattice and the function $\sim$.
   This axiom is to be included only if τ is finite.
3. Axioms for $^\circ$

   $(^\circ{}_{1S})$ $A^\circ \leftrightarrow (\neg A)^\circ$
   $(^\circ{}_{2S})$ $A^\circ \leftrightarrow (f_\lambda A)^\circ$
   $(^\circ{}_{3S})$ $(A^\circ)^\circ$, $(A \vee B)^\circ$, $(A \wedge B)^\circ$, $(A \to B)^\circ$
4. The axioms for annotated variables are replaced by the following axioms:

   (τ_{1S}) $\neg(A^\circ) \to f_\perp A$
   (τ_{2S}) $\neg(A^\circ) \to (\neg^k f_\lambda A \leftrightarrow \neg^{k-1} f_{\sim\lambda} A)$ for $k \geq 1$
   (τ_{3S}) $f_\mu A \to f_\lambda A$ for $\mu \geq \lambda$
   (τ_{4S}) $f_\mu f_\lambda A \leftrightarrow f_\mu A$
   (τ_{5S}) $\neg(A^\circ) \to (f_\lambda A \leftrightarrow f_\lambda \neg A)$
   (τ_{6S}) $A^\circ \to (f_\lambda A \leftrightarrow A)$
5. Rules of Inference

   (R1) Modus Ponens
   $$\frac{A, A \to B}{B}$$
   (R2) If $J \neq \emptyset$ and for all $j \in J$, $\lambda_j \in |\tau|$ and $\lambda = \bigvee_{j \in J} \lambda_j$, then
   $$\frac{A \to f_{\lambda_j} B \text{ for } j \in J}{A \to f_\lambda B}$$

$SP\tau$ are structural since all axioms and the two rules of inference are closed under substitution. The axiomatization of $P\tau$ has no axioms that take into account the specific function $\sim$. In the finitary case, of the above axiomatization, this is considered by axiom τ_{4S}.

Let $\vdash_{SP\tau}$ be the provability relation and Σ be a set of formulas, and A and B formulas. Then, we have:

Theorem 5.10 $\Sigma, A \vdash_{SP\tau} B \Leftrightarrow \Sigma \vdash_{SP\tau} A \to B$.

Theorem 5.11 *Let $\varphi(x_1, x_2, \ldots, x_n)$ be a classical tautology. Then,*

1. $A_1^\circ, A_2^\circ, \ldots, A_n^\circ \vdash_{SP\tau} \varphi(x_1, x_2, \ldots, x_n)$.
2. *If $\varphi(x_1, x_2, \ldots, x_n)$ does not contain negation, then*

   $\vdash_{SP\tau} \varphi(A_1, A_2, \ldots, A_n)$.

Proof Assuming $A_1^\circ, A_2^\circ, \ldots, A_n^\circ$ by *modus ponens*, we obtain all axioms for negation, which together with those for $\rightarrow$, $\wedge$ and $\vee$, are all we need to prove any classical tautology. If negation is not involved, the axioms for binary connectives suffice.

Theorem 5.12 *Let $\varphi(x_1, x_2, \ldots, x_n)$ be a classical tautology. Then,*

1. *If $A_1, A_2, \ldots, A_n$ are complex formulas, then $\models \varphi(A_1, A_2, \ldots, A_n)$*
2. *If $\varphi(x_1, x_2, \ldots, x_n)$ does not contain negation, then for formulas $A_1, A_2, \ldots, A_n$,*

   $\models \varphi(A_1, A_2, \ldots, A_n)$

Proof From Definition 5.10, the valuation for classical connectives is classical. If $A_1, A_2, \ldots, A_n$ are complex formulas, then the valuation for negation is also classical.

Theorem 5.13 shows the soundness of $SP\tau$. Note that $SP\tau$ is not complete.

Theorem 5.13 *If $\Sigma \vdash_{SP\tau} A$, then $\Sigma \models_{SP\tau} A$.*

Proof From Theorem 5.12(2), we do not need to consider the axioms for $\wedge$, $\vee$ and $\rightarrow$ since they are classical tautologies without negations.

Moreover, since $v_I(A^\circ) = 1 \Leftrightarrow$ if A is a complex formula and the consequents of axioms $\neg_{1S}$ and $\neg_{3S}$ are also classical tautologies, these axioms also hold for any valuation.

Other axioms can be verified as follows. For $(\neg_{2S})$, let I be an interpretation. If A is a hyper-literal, then $v_I(A^\circ) = 0$. So,

$$v_I(A^\circ \rightarrow (A \rightarrow (\neg A \rightarrow B))) = 1.$$

If A is a complex formula, then $v_I(A^\circ) = 1$, so

$$\begin{aligned} v_I(A^\circ \rightarrow (A \rightarrow (\neg A \rightarrow B))) = 0 &\Leftrightarrow v_I(A \rightarrow (\neg A \rightarrow B)) = 0 \\ &\Leftrightarrow v_I(A) = 1 \text{ and } v_I(\neg A \rightarrow B) = 0 \\ &\Leftrightarrow v_I(A) = v_I(\neg A) = 1 \text{ and } v_I(B) = 0 \\ &\Leftrightarrow v_I(A) = 1, v_I(A) = 0, \text{ and } v_I(B) = 0 \end{aligned}$$

But this is impossible since A is a complex formula. Thus, axiom $(\neg_{2S})$ is valid.

For $(\neg_{4S})$, from the above remark on $\sim$, we have that $\sim^{T+P+k_1} \lambda = \sim^{T+k_1}$. If $A = \neg^{k_1} f_{\lambda_1} \neg^{k_2} f_{\lambda_2} \ldots \neg^{k_n} f_{\lambda_n} \neg^{k_{n+1}} p$ and I is an interpretation, then we have:

$$\neg^{T+P} A = \neg^{T+P}(\neg^{k_1} f_{\lambda_1} \neg^{k_2} f_{\lambda_2} \ldots \neg^{k_n} f_{\lambda_n} \neg^{k_{n+1}} p)$$

So, we can obtain the following:

$$\begin{aligned} v_I(\neg^{T+P}A) = 1 &\Leftrightarrow I(p) \geq \sim^{T+P+k_1} \lambda_1 \\ &\Leftrightarrow I(p) \geq \sim^{T+k_1} \lambda_1 \\ &\Leftrightarrow v_I(\neg^T A) = 1 \end{aligned}$$

If A is a complex formula, then since P is a multiple of 2, $v_I(\neg^{T+P}A) = v_I(\neg^T A)$. So in any case, for any interpretation I,

$$v_I(\neg^{T+P}A \leftrightarrow \neg^T A) = 1$$

Thus, axiom $(\neg_{4S})$ is valid.

Next, we check the axioms for $\circ$. For $(\circ_{1S})$, $v_I(A^\circ) = 1$ iff A is a complex formula iff $\neg A$ is a complex formula iff $v_I((\neg A)^\circ) = 1$, so

$$v_I(A^\circ \leftrightarrow (\neg A)^\circ) = 1$$

for any interpretation I.

For $(\circ_{2S})$, $v_I(A^\circ) = 1$ iff A is a complex formula iff $f_\lambda A$ is a complex formula iff $v_I((f_\lambda A)^\circ) = 1$, so

$$v_I(A^\circ \leftrightarrow (f_\lambda A)^\circ) = 1$$

for any interpretation I.

For $(\circ_{3S})$, C is $A^\circ, A \wedge B, A \vee B$ or $A \rightarrow B$, since these are all complex formulas, for any interpretation I, $v_I(C^\circ) = 1$.

Finally, we check the axioms for τ. For (τ_{1S}), if $v_I(\neg(A^\circ)) = 1$ then A is a hyper-literal. Here, let p be a propositional variable. Then, since $I(p) \geq \bot$, we have that $v_I(f_\bot A) = 1$. Thus, for any interpretation I,

$$v_I(\neg(A^\circ) \rightarrow f_\bot A) = 1.$$

For (τ_{2S}), if $v_I(\neg(A^\circ)) = 1$ then A is a hyper-literal. Thus, $v_I(\neg^k f_\lambda A) = v_I(\neg^{k-1} f_{\sim\lambda} A)$ for any interpretation I, so

$$v_I(\neg(A^\circ) \rightarrow (\neg^k f_\lambda A \leftrightarrow f_{\sim^k\lambda} A)) = 1.$$

For (τ_{3S}), let $\mu \geq \lambda$. If A is a hyper-literal, p is its propositional variable and $v_I(f_\mu A) = 1$, then $I(p) \geq \mu \geq \lambda$ so $v_I(f_\lambda A) = 1$.

If A is a complex formula, then so are $f_\lambda A$ and $f_\mu A$. So we have that $v_I(A) = v_I(f_\lambda A) = v_I(f_\mu A)$. Thus, in both cases, for any interpretation I,

$$v_I(f_\mu A \rightarrow f_\lambda A) = 1.$$

For (τ_{4S}), if A is a hyper-literal and p is its propositional variable. Then, $v_I(f_\mu f_\lambda A) = 1$ iff $I(p) \geq \mu$ iff $v_I(f_\mu A) = 1$.

If A is a complex formula, so is $f_\lambda A$, so

$$v_I(f_\mu f_\lambda A) = v_I(f_\lambda A) = v_I(A) = v_I(f_\mu A)$$

Thus, for any interpretation I, we have:

$$v_I(f_\mu f_\lambda A \leftrightarrow f_\mu A) = 1$$

For (τ_{5S}), if $v_I(\neg(A^\circ) = 1$, then A is a hyper-literal and so are $f_\lambda A$ and $f_\lambda \neg A$. They have the same propositional variable p and $v_I(f_\lambda A) = 1$ iff $I(p) \geq \lambda$ iff $v_I(f_\lambda \neg A) = 1$. Thus, for any interpretation I, we have:

$$v_I(\neg(A^\circ) \rightarrow (f_\lambda A \leftrightarrow f_\lambda \neg A)) = 1$$

For (τ_{6S}), if $v_I(A^\circ) = 1$, then A is a complex formula and so is $f_\lambda A$. So, $v_I(A) = v_I(f_\lambda A)$. Thus, for any interpretation I, we have:

$$v_I(A^\circ \rightarrow (f_\lambda A \leftrightarrow A)) = 1$$

We here check the soundness of rules (R1) and (R2). The soundness of (R1) is obvious. For (R2), let I be a valuation such that for each $j \in J$, $v_I(A \rightarrow f_{\lambda_j} B) = 1$ and $v_I(A \rightarrow f_\lambda B) = 0$. This implies that $v(A) = 1$ and $v_I(f_\lambda B) = 0$.

Suppose B is a hyper-literal and p is its propositional variable. Since $v_I(f_\lambda B) = 0$, we have $I(p) \not\geq \lambda$. But, $v_I(f_{\lambda_i} B) = 1$ for all $j \in J$. Thus, $I(p) \geq \lambda_j$, and we have that $I(p) \geq \bigvee_{j \in J} \lambda_j = \lambda$. But, this is a contradiction.

If B is a complex formula, since $J \neg \emptyset$ and $v_I(A) = 1$ by assumption, there is a $j \in J$ such that $v_I(f_{\lambda_j} B) = 1$, so $v_I(f_\lambda B) = 1$. Thus, we have that $v_I(A \rightarrow f_\lambda B) = 1$.

Form these, if $v_I(A \rightarrow f_{\lambda_j} B) = 1$ for each $j \in J$, then $v_I(A \rightarrow f_\lambda B) = 1$ for any interpretation I. Thus, (R2) is sound.

Now, we show the relationships between $P\tau$ and $SP\tau$. For so doing, we define syntactic translations between these systems.

Definition 5.11 Let $\mathscr{F}_S$ be the set of formulas of $SP\tau$ and $\mathscr{F}$ the set of formula of $P\tau$. A translation $\theta : \mathscr{F}_S \rightarrow \mathscr{F}$ is recursively defined as follows:

1. If A is a hyper-literal, then $\theta(A) = A$.
2. If $A = B^\circ$, then $\theta(A) = \neg(f_\top \theta(B) \leftrightarrow f_\top \theta(\neg B))$.
3. If $A = f_\lambda B$ and B is a complex formula, then $\theta(A) = f \lambda \theta(B)$.
4. If $A = \neg B$ and B is a complex formula, then $\theta(A) = \neg \theta(B)$.
5. If $A = B * C$, then $\theta(A) = \theta(B) * \theta(C)$ for any binary connective.

Observe that $\theta(A)$ does not contain $^\circ$ and that if A does not contain $^\circ$ then $\theta(A) = A$.

Next, we recursively define $\chi : \mathscr{F}_S \rightarrow \mathscr{F}$ as follows:

1. If $A = p$, then $\chi(A) = p_\top$.
2. If $A = f_\lambda \neg^{k_n} f_{\lambda_n} \ldots \neg^{k_1} f_{\lambda_1} \neg^{k_0} p$, then $\chi(A) = p_\lambda$.
3. If $A = f_\lambda B$, for a complex formula B, then $\chi(A) = \chi(B)$
4. If $A = \neg B$, then $\chi(A) = \neg \chi(B)$.
5. If $A = B * C$, then $\chi(A) = \chi(B) * \chi(C)$ for any binary connective $*$.

Finally, we define

$$\psi(A) : \mathscr{F}_S \to \mathscr{F}$$
$$A \mapsto \chi(\theta(A))$$

Similarly, we recursively define $\varphi : \mathscr{F} \to \mathscr{F}_S$ as follows.

1. If $A = p_\lambda$, then $\varphi(A) = f_\lambda p$.
2. $A = \neg B$, then $\varphi(A) = \neg\varphi(B)$.
3. If $A = B * C$, for binary connective $*$, then $\varphi(A) = \varphi(B) * \varphi(C)$

Theorem 5.14 *The following hold.*

1. *If* $\{A_1, \ldots, A_n\} \vdash_{P\tau} A$, *then* $\{\varphi(A_1), \ldots, \varphi(A_n)\} \cup \{\neg(p^\circ) \mid p \in \mathscr{P}\} \vdash_{SP\tau} \varphi(A)$
2. *If* $\{A_1, \ldots, A_n\} \vdash_{SP\tau} A$, *then* $\{\psi(A_1), \ldots, \psi(A_n)\} \vdash_{P\tau} \psi(A)$.

Proof (1) Although for an infinite lattice τ, rule (τ_4) is infinitary, our proof may be infinite, limit stages of the proof can only come from an application of rule (τ_4).

Let $\langle \sigma_1, ., , , ., \sigma \rangle$, where α is an ordinal, be a proof in $P\tau$ of A from $A_1, \ldots, A_n$. We will turn it into a proof of $\varphi(A)$ in $SP\tau$ by replacing each σ_i by a finite set of formulas, that includes $\varphi(\sigma_i)$, all of which are provable in $SP\tau$ from $\{\varphi(A_1), \ldots, \varphi(A_n)\} \cup \{\neg(p^\circ) \mid p \in \mathscr{P}\}$. This will be done by induction on the length of the proof.

Axioms for $\to$, $\wedge$, $\vee$ are obvious since $\varphi(\sigma_i)$ is an instance of the same axioms in $SP\tau$. So we replace σ_i by $\varphi(\sigma_i)$.

If $\sigma_i = (F \to G) \to ((F \to \neg G) \to \neg F)$, where F and G are complex formulas, then

$$\varphi(\sigma_i) = (\varphi(F) \to \varphi(G)) \to ((\varphi(F) \to \neg\varphi(G)) \to \neg\varphi(F))$$

But, since $\varphi(F), \varphi(G)$ are complex formulas,

$$\vdash_{SP\tau} \varphi(F)^\circ \text{ and } \vdash_{SP\tau} \varphi(G)^\circ$$

and these together with axiom $\neg_{1S}$ yield:

$$\vdash_{SP\tau} \varphi(\sigma_i).$$

So, we replace σ_i by the set of formulas

$$\varphi(F)^\circ, \varphi(G)^\circ, (\varphi(F)^\circ \wedge \varphi(G)^\circ) \to \varphi(\sigma_i) \text{ and } \varphi(\sigma_i)$$

of which the first two are theorems, the third is an axiom of $SP\tau$, and the last one is obtained from the others by *modus ponens*.

If $\sigma_i = p_\perp$, then $\varphi(\sigma_i) = f_\perp p$. Then, we can replace σ_i by the set

$$\neg(p^\circ) \to f_\perp p, \neg(p^\circ) \text{ and } \varphi(\sigma_i)$$

of which the first is an axiom of $SP\tau$, the second is in $\{\neg(p)^\circ \mid p \in \mathscr{P}\}$ and the last one is obtained from the others by *modus ponens*.

If $\sigma_i = \neg^k p_\mu \leftrightarrow \neg^{k-1} p_{\sim\mu}$, then

$$\varphi(\sigma_i) = \neg^k f_\mu p \leftrightarrow \neg^{k-1} f_{\sim\mu} p.$$

We replace by

$$\neg(p^\circ) \to (\neg^k f_\mu p \leftrightarrow \neg^{k-1} f_{\sim\mu} p),\ \neg(p^\circ) \text{ and } \varphi(\sigma_i)$$

of which the first is an axiom of $SP\tau$, the second is in $\{\neg(p)^\circ \mid p \in \mathscr{P}\}$ and the last one is obtained from the others by *modus ponens*.

If $\sigma_i = p_\mu \to p_\lambda$, for $\mu \geq \lambda$, then

$$\varphi(\sigma_i) = f_\mu p \to f_\lambda p$$

is an axiom of $SP\tau$.

If $\sigma_i = A \to p_\mu$, where $\mu = \bigvee_{j \in J} \mu_j$, and σ_i is obtained by rule τ_4 from $\sigma_j = A \to p_{\mu_j}$, where for all $j \in J, j < i$. If we assume $J \neq \emptyset$, then

$$\varphi(\sigma_i) = \varphi(A) \to f_\mu p$$

and for all $j \in J$,

$$\varphi(\sigma_j) = \varphi(A) \to f_{\mu_j} p$$

Then, using rule τ_{7S}, we have $\varphi(\sigma_i)$.

Finally, if σ_i is obtained by *modus ponens* from a pair of premises, since this is a rule in both systems, $\varphi(\sigma_i)$ is also obtained from the corresponding premises.

This completes the proof of (1).

(2) Let $\langle \sigma_1, \ldots, \sigma_\alpha \rangle$, where α is an ordinal, be a proof of A in $SP\tau$ from $A_1, \ldots, A_n$. We prove by induction on the length of the proof that for all $i \leq \alpha$,

$$\{\psi(A_1), \ldots, \psi(A_n)\} \vdash_{P\tau} \psi(\sigma_i).$$

We need to check that if σ is an axiom of $SP\tau$ or it is obtained from previous σ_j's by one of the two rules, then $\psi(\sigma_i)$ is a theorem of $P\tau$.

Since for axioms for $\to$, $\wedge$ and $\vee$, $\psi(\sigma_i)$ is an instance of the same axioms in $P\tau$, the assertion holds for all these axioms. The same can be said if σ_i is obtained by *modus ponens* from previous premises.

Now, we look at other axioms. Observe that A is a complex formula in $SP\tau$ iff $\theta(A)$ is a complex formulas in $SP\tau$ iff $\psi(A)$ is a complex formula in $P\tau$ and that

$$\begin{aligned}
\psi(A^\circ) &= \chi(\neg(f_\top \theta(A) \leftrightarrow f_\top \theta(\neg A))) \\
&= \neg(\chi(f_\top \theta(A)) \leftrightarrow \chi(f_\top \neg\theta(A))) \\
&= \neg(\chi(\theta(A)) \leftrightarrow \neg\chi(\theta(A))) \\
&= \neg(\psi(A) \leftrightarrow \neg\psi(A))
\end{aligned}$$

and also

$$\psi(\neg(A)^\circ) = \neg\neg(\psi(A) \leftrightarrow \neg\psi(A)),$$

so if A is a complex formula, the following hold.

$\vdash_{P\tau} \psi(A^\circ), \vdash_{P\tau} \neg\psi(\neg(A^\circ))$.

On the other hand, is A is a hyper-literal, assuming

$$A = \neg^{k_n} f_{\lambda_n} \neg^{k_{n-1}} f_{\lambda_{n-1}} \ldots \neg^{k_1} f_{\lambda_1} \neg^{k_0} p,$$

then

$$\theta(A) = A,$$

and

$$\begin{aligned}\psi(A^\circ) &= \neg(\chi(f_\top \theta(A) \leftrightarrow f_\top \theta(\neg A)))\\ &= \neg(\chi(f_\top A) \leftrightarrow \chi(f_\top \neg A))\\ &= \neg(p_\top \leftrightarrow p_\top)\end{aligned}$$

and also

$$\psi(\neg(A^\circ)) = \neg\neg(\psi(A) \leftrightarrow \neg\psi(A))$$

so if A is a complex formula, then the following hold.

$\vdash_{P\tau} \psi(A^\circ)$
$\vdash_{P\tau} \neg\psi(\neg(A^\circ))$.

On the other hand, if A is a hyper-literal, assuming

$$A = \neg^{k_n} f_{\lambda_n} \neg^{k-1} f_{\lambda_{n-1}} \ldots \neg^{k_1} f_{\lambda_1} \neg^{k_0} p$$

then

$$\theta(A) = A$$

and

$$\begin{aligned}\psi(A^\circ) &= \neg(\chi(f_\top \theta(A) \leftrightarrow f_\top \theta(\neg A)))\\ &= \neg(\chi(f_\top A) \leftrightarrow \chi(f_\top \neg A))\\ &= \neg(p_\top \leftrightarrow p_\top)\end{aligned}$$

and also

$$\chi(\neg(A^\circ)) = \neg\neg(\psi(A) \leftrightarrow \neg\psi(A)).$$

So if A is a hyper-literal, the following hold.

$\vdash_{P\tau} \neg\psi(A^\circ)$
$\vdash_{P\tau} \psi(\neg(A^\circ))$.

So let us now assume σ_i is the axiom of negation $\neg_{1S}$,

$$(A^\circ \wedge B^\circ) \rightarrow ((A \rightarrow B) \rightarrow ((A \rightarrow \neg B) \rightarrow \neg A))$$

then $\psi(\sigma_i)$ equals

$$(\psi(A^\circ) \wedge \psi(B^\circ)) \rightarrow ((\psi(A) \rightarrow \psi(B)) \rightarrow ((\psi(A) \rightarrow \neg\psi(B)) \rightarrow \neg\psi(A))).$$

We have two cases.

If either A or B is hyper-literal, then using $\vdash_{P\tau} \neg\psi(A^\circ)$, the fact that $\psi(\neg(A^\circ))$ and $\psi(\neg(B^\circ))$ are complex formulas in $P\tau$ and a couple of classical tautologies, we have:

$$\vdash_{P\tau} \psi(\sigma_i)$$

If both A and B are complex formulas in $SP\tau$, then both $\psi(A)$ and $\psi(B)$ are complex formulas in $P\tau$, so

$$(\psi(A) \rightarrow \psi(B)) \rightarrow ((\psi(A) \rightarrow \neg\psi(B)) \rightarrow \neg\psi(A))$$

is an instance of axiom $\neg_1$ and by the deduction theorem (Theorem 2.8) we also have

$$\vdash_{P\tau} \psi(\sigma_i)$$

The next two axioms for negation are treated similarly. In the finitary case, axiom $\neg_{4S}$ needs to be handled more carefully since it has no analog in $P\tau$.

If σ_i is an instance of $\neg_{4S}$, $\neg^{T+P}A \leftrightarrow \neg^T A$, then one can easily check that

$$\psi(\sigma_i) = \neg^{T+P}\psi(A) \leftrightarrow \neg^T\psi(!)$$

Let $I : \mathscr{P} \rightarrow \mid\tau\mid$ be an interpretation. If $A = \neg^{k_1} f_{\lambda_1} \neg^{k_2} f_{\lambda_2} \ldots \neg^{k_n} f_{\lambda_n} \neg^{k_{n+1}} p$, then $\psi(A) = \neg^{k_1} p_{\lambda_1}$. Since $\sim^{T+P+k_1} \lambda_1 = \sim^{T+k_1} \lambda_1$ as stated above, we have:

$$v_I(\neg^{T+P}\psi(A)) = 1 \Leftrightarrow v_I(\neg^T\psi(A)) = 1$$

So we can apply Theorem 2.17 (completeness theorem for $P\tau$ with finite lattice τ).

If A is a complex formula, then $v_I(\neg^{T+P}\psi(A) = v_I(\neg^T\psi(A))$, since P is a multiple of 2. In any case, $\models_{P\tau} \psi(\sigma_i)$ and so by Theorem 2.17, we have:

$$\vdash_{P\tau} \psi(\sigma_i)$$

If σ_i is one of the axioms $\circ_{1S}$ or $\circ_{2S}$, for any formula A, we have that $\psi(A)$ i a complex formula iff $\psi(f_\lambda A)$ is a complex formula iff $\psi(\neg A)$ is a complex formula, then by $\vdash_{P\tau} \psi(A^\circ)$ and $\vdash_{P\tau} \neg\psi(A^\circ)$ above, $\psi(\sigma_i)$ is a theorem of $P\tau$.

If σ_i is one of the instances of axioms $\circ_{3S}$, by $\vdash_{P\tau} \neg\psi(\neg A^\circ)$, $\psi(\sigma_i)$ is also a theorem of $P\tau$. Thus, in any case, if σ_i is an axiom for $\circ$, then

$$\vdash_{P\tau} \psi(\sigma_i).$$

We now check the axiom τ_S. If σ_i is τ_{1S}, $A^\circ \rightarrow f_\perp A$ and A is a hyper-literal, then

$$\psi(\sigma_i) = \psi(A^\circ) \rightarrow p_\perp.$$

But $p_\perp$ is an axiom of $P\tau$, so by the deduction theorem, $\psi(\sigma_i)$ is a theorem of $P\tau$.

If A is a complex formula, then

$$\psi(\sigma_i) = \psi(A^\circ)\psi(A)$$

so by $\vdash_{P\tau} \psi(A^\circ)$, and the fact that $\psi(A^\circ)$ is a complex formula, after using of couple of classical tautologies, $\psi(\sigma_i)$ is also a theorem $P\tau$.

If σ_i is τ_{2S}, $\neg(A^\circ) \to (\neg^k f_\lambda A \leftrightarrow f_{\sim^k\lambda} A)$ holds. If A is a hyper-literal with propositional variable p, we have

$$\psi(\neg^k f_\lambda A \leftrightarrow f_{\sim^k\lambda} A = \neg^k p_\lambda \leftrightarrow p_{\sim^k\lambda}.$$

and the latter is a theorem of $P\tau$, so by the deduction theorem, $\psi(\sigma_i)$ is also a theorem.

If A is a complex formula, then

$$\psi(\sigma_i) = \sigma(\neg(A^\circ)) \to \psi(\neg^k f_\lambda A \leftrightarrow f_{\sim^k\lambda} A)$$

Since both the antecedent and consequent are complex formulas, using $\vdash_{P\tau} \neg\psi(\neg(A^\circ))$, and several classical tautologies, we get that $\psi(\sigma_i)$ is a theorem $P\tau$.

Similarly, if σ_i is τ_{3S}, then $f_\mu A \to f_\lambda A$ for $\mu \geq \lambda$. If A is a complex formula, then

$$\psi(\sigma_i) = \psi(A) \to \psi(A)$$

If A is a hyper-literal, then

$$\psi(\sigma) = p_\mu \to p_\lambda$$

So in any case

$$\vdash_{P\tau} \psi(\sigma_i).$$

If σ_i is τ_{4S}, $f_\mu f_\lambda \leftrightarrow f_\mu A$ and A is a complex formula, then

$$\psi(\sigma_i) = \psi(A) \leftrightarrow \psi(A).$$

If A is a hyper-literal, then

$$\psi(\sigma_i) = p_\mu \leftrightarrow p_\mu.$$

So in any case

$$\vdash_{P\tau} \psi(\sigma_i).$$

Axioms τ_{5s} and τ_{6s} are led in the same way.

Finally, assume $\sigma_i = A \to f_\lambda B$ is obtained by an application of rule τ_{7s} from $A \to f_{\lambda_j} B, j < i$ for all $j \in J$, where $J \neq \emptyset$ and $\bigvee_{j \in J} \lambda_i$.

If B is a complex formula, then for all $j \in J$,

$$\psi(\sigma_i) = \psi(A \to f_\lambda B) = \psi(A) \to \psi(B) = \psi(A \to f_{\lambda_j} B),$$

so by inductive hypothesis $\psi(\sigma_i)$ is a theorem of $P\tau$.

If B is a hyper-literal,

$$\psi(A \to f_{\lambda_j} B) = \psi(A) \to p_{\lambda_j}$$

and thus

$$\psi(A \to f_{\lambda_j} B) = \psi(A) \to p_\lambda$$

is obtained by an application of rule τ_4.

This ends the proof of the theorem.

The main result of Lewin et al. [109] is that a system $SP\tau$ is *algebraizable* iff τ is finite, although $P\tau$ is not algebraizable.

When τ is infinite, the system cannot be algebraizable in the sense of *abstract logic* due to Blok and Pigozzi [54] since we need an infinitary rule. But, their method can be generalized for $SP\tau$.

For proving that $SP\tau$ is algebraizable, we need to find unary terms $\delta_i, \varepsilon_i, i \in I$ and binary terms $\Delta_j, j \in J$, where I and J are finite, such that

(i) $\vdash_{SP\tau} A\Delta A$

(ii) $A\Delta B \vdash_{SP\tau} B\Delta A$

(iii) $A\Delta B, B\Delta A \vdash_{SP\tau} A\Delta A$

(iv) (a) $A\Delta B \vdash_{SP\tau} A^\circ \Delta B^\circ$
(b) $A\Delta B \vdash_{SP\tau} \neg A\Delta\neg B$
(c) For each $\mu \in \mid \tau \mid, A\Delta B \vdash_{SP\tau} f_\mu A\Delta f_\mu B$
(d) $A_1\Delta B_1, A_2\Delta B_2 \vdash_{SP\tau} A_1 \wedge A_2\Delta B_1 \wedge B_2$
(e) $A_1\Delta B_1, A_2\Delta B_2 \vdash_{SP\tau} A_1 \vee A_2\Delta B_1 \vee B_2$
(f) $A_1\Delta B_1, A_2\Delta B_2 \vdash_{SP\tau} A_1 \rightarrow A_2\Delta B_1 \rightarrow B_2$

(v) $A \dashv\vdash \delta(A)\Delta\varepsilon(A)$

Here, $\vdash_{SP\tau} A\Delta B$ means that for all $j \in J$, $\vdash_{SP\tau} \Delta_j(A, B)$ and similarly for δ and ε. $A \dashv\vdash B$ is shorthand for $A \vdash B$ and $B \vdash A$.

Δ is called a *system of equivalence formulas for S*, and $\delta \approx \varepsilon$ are called *defining equations*.

Theorem 5.15 reveals that a certain set of formulas of $SP\tau$ define a congruence on $\mathcal{F}_S$ that verifies the above conditions, that is $Sp\tau$ is weakly congruential. In general, $SP\tau$ is not algebraizable since this set might be infinite.

Theorem 5.15 *For any lattice τ, let*

$$\delta(A) = A \wedge A$$
$$\varepsilon(A) = A \rightarrow A$$

Define equivalence formulas of three types,

(a) $\Delta_0 \models A^\circ \leftrightarrow B^\circ$

(b) $\Delta_0(A, B) = A \leftrightarrow B,$

$\vdots$

$\Delta_k(, B) = \neg^k A \leftrightarrow \neg^k B,$

$\vdots$

(c) $\Delta_\lambda(A, B) = f_\lambda A \leftrightarrow f_\lambda B,$

for each $\lambda \in \tau$.

Then, $\Delta = \{\Delta_j \mid j \in \{o\} \cup \omega \cup \tau\}$, and the single defining equation $\delta \approx \varepsilon$ (i.e. $\mid I \mid= 1$), verify conditions (i) through (v) above.

Proof The proofs of (i), (ii) and (iii) above are immediate.
For a proof of (iv)(a), by axiom $\circ_{3S}$,

$$\vdash (A^\circ)^\circ \text{ and } \vdash (B^\circ)^\circ,$$

so we have:

$$\vdash (A^\circ)^\circ \leftrightarrow (B^\circ)^\circ,$$

that is,

$$\vdash \Delta_\circ(A^\circ, B^\circ).$$

By Theorem 5.11, for all n,

$$(A^\circ)^\circ, (B^\circ)^\circ \vdash (\neg^n(A^\circ) \leftrightarrow \neg^n(B^\circ)) \leftrightarrow (A^\circ \leftrightarrow B^\circ),$$

since the latter is a classical tautology. So

$$\Delta_\circ(A, B) \vdash \Delta_n(A^\circ, B^\circ).$$

Finally, by axiom τ_{6S},

$$(A^\circ)^\circ \vdash f_\lambda(A^\circ) \leftrightarrow A^\circ,$$
$$(B^\circ)^\circ \vdash f_\lambda(B^\circ) \leftrightarrow B^\circ,$$

so we have:

$$\Delta_\circ(A, B) \vdash \Delta_\lambda(A^\circ, B^\circ)$$

This completes the proof of

$$A \Delta B \vdash A^\circ \Delta B^\circ$$

To prove (iv)(b), by axiom $\circ_{1S}$

$$A^\circ \leftrightarrow B^\circ \vdash (\neg A)^\circ \leftrightarrow (\neg B)^\circ.$$

So

$$A \Delta B \vdash \Delta_\circ(\neg A, \neg B)$$

It is immediate that

$$A \Delta B \vdash \Delta_n(\neg A, \neg B)$$

By axiom τ_{5S},

$$\neg(A^\circ) \vdash f_\lambda A \leftrightarrow f_\lambda A,$$
$$\neg(B^\circ) \vdash f_\lambda B \leftrightarrow f_\lambda B, \text{ so}$$
$$\Delta_\lambda(A, B), \neg(A^\circ), \neg(A^\circ) \vdash f_\lambda \neg A \leftrightarrow f_\lambda \neg B. \quad (5.1)$$

Also, by axiom τ_{6S},

$$A^{\circ} \vdash f_{\lambda}\neg A \leftrightarrow \neg A,$$
$$B^{\circ} \vdash f_{\lambda}\neg B \leftrightarrow \neg B,$$
$$\Delta_1(A, B), A^{\circ}, A^{\circ} \vdash f_{\lambda}\neg A \leftrightarrow f_{\lambda}\neg B. \quad (5.2)$$

So from (5.1) and (5.2), using a classical argument, we prove

$$A\Delta B \vdash f_{\lambda}A \leftrightarrow f_{\lambda}\neg B.$$

This ends the proof of (iv)(b).

Next, we prove (iv)(c). Let $\mu \in |\tau|$. By axiom τ_{2S}

$$A^{\circ} \leftrightarrow B^{\circ} \vdash (f_{\mu}A)^{\circ} \leftrightarrow (f_{\mu}B)^{\circ},$$

So

$$A\Delta B \vdash \Delta_{\circ}(f_{\mu}A, f_{\mu}B).$$

To check the conditions for negation, we will use the same method as in (iv)(b).

By axiom τ_{2S},

$$\neg(A^{\circ}) \vdash \neg^n f_{\mu}A \leftrightarrow f_{\sim^n\mu}A),$$
$$\neg(B^{\circ}) \vdash \neg^n f_{\mu}A \leftrightarrow f_{\sim^n\mu}B),$$
$$\Delta_{\sim^n\mu}(A, B), \neg(A^{\circ}), \neg(B^{\circ}) \vdash \neg^n f_{\mu}A \leftrightarrow \neg^n f_{\mu}B. \quad (5.3)$$

On the other hand. by axiom τ_{6S}, and a couple of classical tautologics,

$$A^{\circ} \vdash \neg^n f_{\mu}A \leftrightarrow \neg^n A,$$
$$B^{\circ} \vdash \neg^n f_{\mu}B \leftrightarrow \neg^n B,$$
$$\Delta_{\sim^n\mu}(A, B), A^{\circ}, B^{\circ} \vdash \neg^n f_{\mu}A \leftrightarrow \neg^n f_{\mu}B. \quad (5.4)$$

From (5.3) and (5.4), we have:

$$A\Delta B \vdash \Delta_n(f_{\mu}A, f_{\mu}B).$$

Next, by axiom τ_{4S},

$$\Delta_{\lambda}(A, B) \vdash f_{\lambda}f_{\mu}A \leftrightarrow f_{\lambda}f_{\mu}B,$$

which completes the proof that

$$A\Delta B \vdash f_{\mu}A\Delta f_{\mu}B.$$

The proofs of (iv)(d), (e) and (f) follow from the fact that if A and B behave classically, then it is straightforward to check that

$$A \leftrightarrow B \vdash A\Delta B.$$

If $*$ is a binary connective, then $A_1 * A_2$ and $B_1 * B_2$ are complex formulas, so it is enough to prove that

$$A_1 \Delta B_1, A_2 \Delta B_2 \vdash (A_1 * A_2) \leftrightarrow (B_1 * B_2),$$

which in all three cases can be obtained using *modus ponens* and a classical tautology that does not contain a negation.

For a proof of (v), since $\delta(A)$ and $\varepsilon(A)$ are complex formulas, it is enough to prove

$$A \vdash \delta(A) \leftrightarrow \varepsilon(A).$$

Since $\vdash A \rightarrow A$, we have that $\vdash (A \wedge A) \rightarrow (A \rightarrow A)$. On the other hand, $A \vdash A \wedge A$, so $A \vdash (A \rightarrow A) \rightarrow (A \wedge A)$, that is $A \vdash (A \wedge A) \leftrightarrow (A \rightarrow A)$.

Also, $(A \wedge A)\Delta(A \rightarrow A) \vdash (A \rightarrow A) \rightarrow (A \wedge A)$, but since $\vdash A \rightarrow A$, and $A \wedge A \vdash A$, $(A \wedge A)\Delta(A \rightarrow A) \vdash A$.

If τ is finite, then $SP\tau$ are algebraizable, but this is not true in general. Now, we prove that $SP\tau$ are not algebraizable for an infinite lattice τ, which can be proved by showing that for a given lattice, the Leibniz equality function $\boldsymbol{\Omega}$ defined in [54] does not preserve unions of directed subsets of $Th(SP\tau)$.

Lemma 5.1 *Let* $\mathbf{A}$ *be an algebra and* $F \subseteq A$. *Let* Θ *be a binary operation on* A *that is definable by a set of formulas over the matrix* $\langle \mathbf{A}, F\rangle$ *with parameters and without equality.*

1. *If* Θ *is reflexive, then* $\boldsymbol{\Omega}_{\mathbf{A}}F \subseteq \Theta$.
2. *If, in addition,* Θ *is a congruence on* $\mathbf{A}$, *that is compatible with* F, *then* $\boldsymbol{\Omega}_{\mathbf{A}}F = \Theta$.

Proof Let Θ be defined by an arbitrary set of formulas. Then, since Θ is reflexive, for any $\gamma[p, q, r_0, \ldots, r_{k-1}]$ with parameters $c_0, \ldots, c_{k-1}$ in the defining set and any $a \in A$,

$$\langle \mathbf{A}, F\rangle \models \gamma[p, q, r_0, \ldots, r_{k-1}]$$

Now, let $(a, b) \in \boldsymbol{\Omega}_{\mathbf{A}}F$, then by the definition of the Leibniz relation, for any such γ and set of parameters,

$$\langle \mathbf{A}, F\rangle \models \gamma[p, q, r_0, \ldots, r_{k-1}]$$

and thus

$$\boldsymbol{\Omega}_{\mathbf{A}}F \subseteq \Theta.$$

For proving (ii), we simply recall that $\boldsymbol{\Omega}_{\mathbf{A}}F$ is the largest congruence of A that is compatible with F, so $\Theta \subseteq \boldsymbol{\Omega}_{\mathbf{A}}F$.

Lemma 5.2 *Let* T *be a theory in* $SP\tau$ *and define*

$$\Theta = \{(\varphi, \psi) \mid \varphi\Delta\psi \in T\}.$$

Then $\boldsymbol{\Omega}T = \Theta$.

Proof From [54], we recall that $\boldsymbol{\Omega} T$ is the largest congruence over $\mathscr{F}$ compatible with T.

By Theorem 5.15, Θ is a congruence.

It is obvious that Θ is compatible with T, for if $(\varphi, \psi) \in \Theta$ and $\varphi \in T$, then $\varphi \Delta \psi \in T$, so in particular, $\varphi \leftrightarrow \psi \in T$ and by *modus ponens*, $\psi \in T$.

Finally, since Θ is reflexive and defined by a set of formulas, by Lemma 5.1, $\boldsymbol{\Omega} T = \Theta$.

Lemma 5.3 *Let $\Delta \subseteq \tau$ be such that $\perp \in \Delta$. Then*

$$\{f_\gamma p \leftrightarrow f_\gamma q \mid \gamma \in \Delta\} \vdash f_\beta p \leftrightarrow f_\beta q$$

iff there is a subset Δ of τ, such that

$$\beta = \bigvee_{\lambda \in \Lambda} \lambda.$$

Proof Let

$$\beta = \bigvee_{\lambda \in \Lambda} \lambda.$$

If $\Lambda = \emptyset$, then $\beta = \perp \in \Lambda$, and the theorem holds.

Assume $\Delta \neq \emptyset$. Now, since $\beta \geq \lambda$ for each $\lambda \in \Lambda$, by axiom τ_{3S}, $\vdash f_\beta p \rightarrow f_\lambda p$, so

$$\{f_\gamma p \leftrightarrow f_\gamma q \mid \gamma \in \Delta\} \vdash f_\beta p \rightarrow f \lambda q,$$

for each $\lambda \in \Delta$. So by rule τ_{7S},

$$\{f_\gamma p \leftrightarrow f_\gamma q \mid \gamma \in \Lambda\} \vdash f_\beta p \rightarrow f_\beta q.$$

Similarly, we get the other direction, so

$$\{f_\gamma p \leftrightarrow f_\gamma q \mid \gamma \subset \Lambda\} \vdash f_\beta p \leftrightarrow f_\beta q.$$

Now, assume that β is not the supremum of any subset of Λ. Let

$$D = \{\alpha \Lambda \mid \alpha \leq \beta\}.$$

Let $\delta = \bigvee D$. Then

$$\delta < \beta.$$

It is also clear that for $\gamma \in \Lambda$,

$$\gamma < \beta \Leftrightarrow \gamma \leq \delta.$$

Let $I(p) = \delta$ and $I(q) = \beta$. The for all $\gamma \in \Lambda$,

$$v_I(f_\gamma p) = 1 \Leftrightarrow v_I(f_\gamma q) = 1$$

and thus,

$$v_I(f_\gamma p \leftrightarrow f_\gamma q) = 1$$

$v_I(f_\beta p) = 0$ and $v_I(f_\beta q) = 1$, so by the completeness theorem of $SP\tau$, the proof is complete.

Theorem 5.16 is a main result about structural annotated logics $SP\tau$.

Theorem 5.16 *An annotated logic $SP\tau$ is algebraizable iff $|\tau|$ is finite.*

Proof If τ is finite, then $\Delta'_\lambda s$ are also finite. Also, b axiom $\neg_{4S}$, it is enough to consider Δ_i for $0 \leq i < T + P$, since if $n \geq T + P$, then $n = T + m + kP$ for $m < P$ and k so

$$\vdash \neg^n A \leftrightarrow \neg^{T+m} A$$

and thus

$$\Delta_{T+m}(A, B) \vdash \Delta_n(A, B).$$

So we have finitely many equivalence formulas and $SP\tau$ is algebraizable by Theorem 5.15.

If τ is infinite, then we will say that a subset X *generates* τ if all elements of τ are supremum of elements of X.

Let Λ be a subset that generates τ of minimal cardinality, in the sense that there is no subset of τ of less cardinality than that of Λ that generates τ. Assume that

$$\Lambda = \{\lambda_i \mid i < \kappa\}$$

For $\gamma < \kappa$, let

$$T_\gamma = Cn(\{f_{\lambda_i} p \leftrightarrow f_{\lambda_i} q \mid i < \gamma\}) \cup \{\neg(p^\circ), \neg(q^\circ)\}$$

and

$$T = Cn \bigcup_{\gamma < \kappa} T_\gamma.$$

Clearly,

$$T \vdash \Delta_0(p, q),$$

so by axiom $\circ_{2S}$,

$$T \vdash \Delta_0(f_\top p, f_\top q),$$

Since Λ generates τ, by Lemma 5.3, for all $\lambda \in |\tau|$,

$$T \vdash f_\lambda p \leftrightarrow_\lambda q,$$

so by axiom τ_{4S},

$$T \vdash f_\lambda(f_\top p) \leftrightarrow f_\lambda(f_\top q),$$

that is

$$T \vdash \Delta_\lambda(f_\top p, f_\top q).$$

Also, since

$$\neg((f_\top p)^\circ) \vdash \neg^n(f_\top p) \leftrightarrow f_{\sim^n\top}(f_\top p)$$

and

$$\neg((f_\top q)^\circ) \vdash \neg^n(f_\top q) \leftrightarrow f_{\sim^n\top}(f_\top q)$$

we have

$$T \vdash \Delta_n(f_\top p, f_\top q)$$

for all $n \geq 0$.

Thus, we have proved that

$$T \vdash f_\top p \Delta f_\top q,$$

so by Lemma 5.2,

$$(f_\top p, f_\top q) \in \boldsymbol{\Omega} T,$$

Now, suppose that

$$(f_\top p, f_\top q) \in \bigcup_{\gamma < \kappa} \boldsymbol{\Omega} T_\gamma.$$

Then for some $\gamma < \kappa$

$$(f_\top p, f_\top q) \in \boldsymbol{\Omega} T_\gamma.$$

But, $\{\lambda_i \mid i < \gamma\}$ has cardinality less than κ, so it does not generate τ by Lemma 5.3,

$$T_\gamma \leq f_\lambda(f_\top p) \leftrightarrow f_\lambda q$$

for some $\lambda \in |\tau|$. But then

$$T_\gamma \leq f_\lambda p \leftrightarrow f_\lambda q,$$

so

$$T_\gamma \leq f \top p \Delta f_\top q.$$

Thus, we get a contradiction.

This proves that $\boldsymbol{\Omega}$ is not preserve directed sets of the theories, so $SP\tau$ is not algebraizable for infinite τ.

Later, Lewin et al. [110] introduced structural annotated logics SAL_τ to improve $SP\tau$ with a matrix semantics. As shown above, $SP\tau$ is equivalent to $P\tau$, and is algebraizable, but we need several unary operation symbols and their axioms.

To overcome difficulties with $P\tau$, Lewin, Mikenberg and Schwarze proposed new structural annotated logics SAL_τ. Below we overview SAL_τ.

Definition 5.12 (*Symbols*) The symbols of SAL_τ are defined as follows:

1. Propositional symbols: $p, q, \ldots$ (possibly with subscript)
2. Annotated constants: $\mu, \lambda, \ldots \in |\tau|$
3. Logical connectives: $\wedge$ (conjunction), $\vee$ (disjunction), $\rightarrow$ (implication), $\neg$ (negation), $^\circ$, and f_λ for each $\lambda \in |\tau|$.
4. Parentheses: (and)

Definition 5.13 (*Formulas*) Formulas are defined as follows:

1. Propositional symbols $p, q, r, \ldots$ are formulas.
2. If A is a formula, then A° is a formula.
3. If A is a formula and λ is an annotation constant, then $f_\lambda A$ is a formula.
4. If A and B are formulas, then $\neg A, A \wedge B, A \vee B$ and $A \rightarrow B$ are formulas.
5. If p is a propositional symbol and $\lambda_i \in |\tau|$ $(0 \leq i \leq n, k_i \in \omega$ (the set of natural numbers)), then $\neg^{k_n} f_{\lambda_n} \neg^{k_{n-1}} f_{\lambda_{n-1}} \cdots \neg^{k_1} f_{\lambda_1} \neg^{k_0} p$ is a formula called a hyper-literal. A formula that is not a hyper-literal is called a complex formula.

SAL_τ use the same symbols and language of SP_τ. Thus, their conventions are similarly interpreted. A formula $f\lambda$ reads "the proposition p has credibility greater than or equal to λ". A formula p° reads "p is well behaved with regard to negation". The differences lie in the axiomatization and semantics.

Postulates for SAL_τ

1. Axioms for binary connectives:

   $(\rightarrow_1)\ p \rightarrow (q \rightarrow p)$
   $(\rightarrow_2)\ (p \rightarrow (q \rightarrow r)) \rightarrow ((p \rightarrow q) \rightarrow (p \rightarrow r))$
   $(\rightarrow_3)\ (p \rightarrow q) \rightarrow p) \rightarrow p$
   $(\wedge_1)\ (p \wedge q) \rightarrow p$
   $(\wedge_2)\ (p \wedge q) \rightarrow q$
   $(\wedge_3)\ p \rightarrow (q \rightarrow (p \wedge q))$
   $(\vee_1)\ p \rightarrow (p \vee p)$
   $(\vee_2)\ q \rightarrow (p \vee q)$
   $(\vee_3)\ (p \rightarrow r) \rightarrow ((q \rightarrow r)\ to\ ((p \vee q) \rightarrow r))$

2. Axioms for negation:

   $(\neg_1)\ (p^\circ \wedge q^\circ) \rightarrow ((p \rightarrow q) \rightarrow ((p \rightarrow \neg) \rightarrow \neg p))$
   $(\neg_2)\ p^\circ \rightarrow (p \rightarrow (\neg p \rightarrow q))$
   $(\neg_3)\ p^\circ \rightarrow (p \vee \neg p)$
   $(\neg_4)\ \neg^{t+s} p \leftrightarrow \neg^t p$

   where t and s are the numbers.

3. Axioms for $^\circ$

   $(\circ_1)\ p^\circ \leftrightarrow (\neg p^\circ)$
   $(\circ_2)\ p^\circ \leftrightarrow (f_\lambda p)^\circ$
   $(\circ_3)\ (p^\circ)^\circ,\ (p \wedge q)^\circ,\ (p \vee q)^\circ,\ (p \rightarrow q)^\circ$

4. Axioms for annotated formulas:

   $(\tau_1)\ \neg(p)^\circ \to f_\perp p$
   $(\tau_2)\ \neg(p)^\circ \to (\neg^k f_\lambda p \leftrightarrow \neg^{k-1} f_{\sim\lambda} p)$, where $k \geq 1$
   $(\tau_3)\ f_\mu p \to f_\lambda p$ for $\mu \geq \lambda$
   $(\tau_4)\ f_\mu f_\lambda p \to f_\mu p$
   $(\tau_5)\ \neg(p^\circ) \to (f_\lambda p \leftrightarrow f_\lambda \neg p)$
   $(\tau_6)\ p^\circ \to (f_\lambda p \leftrightarrow p)$

   For the next two axioms we will need the following definitions:

$$S_\kappa(p) = f\kappa p \wedge \bigwedge_{\lambda \not\leq \kappa} \neg(f_\lambda p \wedge f_\lambda p) \qquad \text{if } \kappa \neq \top$$
$$= \neg(p^\circ) \wedge f_\top p \qquad \text{if } \kappa = \top$$
$$T_{(\lambda,\kappa)} p = \bigwedge_{m < t+s, \sim^m \lambda \leq \kappa} \neg^m p \wedge \bigwedge_{m < t+2, \sim^m \lambda \leq \kappa} \neg(\neg^m p \wedge \neg^m p).$$

$$(\tau_7)\ \neg(p^\circ) \to \bigvee_{\kappa \in \tau} S_\kappa(p)$$
$$(\tau_8)\ \neg(p^\circ) \to \bigvee_{\kappa \in \tau} T_{\lambda,\kappa}(p)$$

5. Inference Rules:

   (R1) *Modus Ponens*: $\dfrac{p, p \to q}{q}$

   (R2) $\dfrac{p \to f_\lambda q, p \to f_\mu q}{p \to f_{\lambda \vee \mu} q}$

For each lattice τ, SAL_τ is an axiomatic extension of the corresponding system $SP\tau$ obtained by adding axioms τ_7 and τ_8. In the axiomatization of $SP\tau$, there is only a crude consideration of the specific lattice τ and the function $\sim$ that we are using. Since τ and $\sim$ are arbitrary, these are not complicated.

Intuitively, the formula $S_\kappa(p)$ is true only if the maximum degree of credibility of the formula p is κ. Axiom τ_7 states that every hyper-literal has a certain maximum degree of credibility.

The formula $T_{(\lambda,\kappa)}(p)$ codes some of the finer aspects of the behavior of the negations of the degree of credibility λ with respect to a given degree of credibility κ. Axiom τ_8 states that if a formula has maximum credibility κ, then its negations will behave like some given degree of credibility λ with respect to κ.

Lewin et al. [110] gave a matrix semantics for annotated logics SAL_τ using a special class $\boldsymbol{M}_\tau$ of SAL_τ-matrices. Since τ is finite, all its ideals are principal, that is, if $\mathscr{I}$ is an ideal, then for some $\kappa \in \tau$, $\mathscr{I} = \{\lambda \in \tau \mid \lambda \leq \kappa\}$. This ideal is denoted by $\mathscr{I}_\kappa$.

For each $I \subseteq \omega$ and $\mathbf{0}, \mathbf{1} \notin \tau$, we define

$$L = (I \times \tau) \cup \{\mathbf{0}, \mathbf{1}\}.$$

and for any I-indexed family $\langle \kappa_i \mid i \in I \rangle$ of elements of τ, let

$$D = \bigcup_{i \in I} (\{i\} \times \mathscr{I}_{k_i} \cup \{\mathbf{1}\}.$$

The set $\boldsymbol{M}_\tau$ of *nice* matrices is

$$\boldsymbol{M}_\tau = \{\mathscr{M} = \langle \boldsymbol{L}, D \rangle \mid I \subseteq \omega, \langle \kappa_i \mid i \in I \rangle \in \tau^I\}$$

where L and D are define as above and the operation of $\boldsymbol{L}$ are defined as follows:

$$a \wedge b = \begin{cases} \mathbf{1} \text{ if } a \in D \text{ and } b \in D, \\ \mathbf{0} \text{ otherwise.} \end{cases}$$

$$a \vee b = \begin{cases} \mathbf{1} \text{ if } a \in D \text{ or } b \in D, \\ \mathbf{0} \text{ otherwise.} \end{cases}$$

$$a \to b = \begin{cases} \mathbf{1} \text{ if } a \notin D \text{ or } b \in D, \\ \mathbf{0} \text{ otherwise.} \end{cases}$$

$$a^\circ = \begin{cases} \mathbf{0} \text{ if } a \in I \times \tau, \\ \mathbf{1} \text{ if } a \in \{\mathbf{0}, \mathbf{1}\}. \end{cases}$$

$$f_\lambda a = \begin{cases} \langle i, \lambda \rangle \text{ if } a = \langle i, \mu \rangle, \\ a \quad\quad \text{ if } a \in \{\mathbf{0}, \mathbf{1}\}. \end{cases}$$

$$\neg a = \begin{cases} \mathbf{0} & \text{if } a = \mathbf{1} \\ \mathbf{1} & \text{if } a = \mathbf{0} \\ \langle i, \sim \mu \rangle & \text{if } a = \langle i, \mu \rangle. \end{cases}$$

A special case of these matrices is when $I = \{i\}$ is a singleton. In this case, we simply drop the ordered pairs and identify $\{i\} \times \tau$ with τ. We call these *elementary* matrices.

SAL_τ can be shown to be complete for a matrix semantics. In [110], reduced matrices were also defined for studying algebraization. Using the matrices, they proved that SAL_τ is a proper quasi-variety as one of the main results.

Lewin, Mikenberg and Schware's works on substructural annotated logics are very important to foundations for annotated logics, in particular, algebraic aspects. It would be interesting to investigate predicate substructural annotated logics.

However, from a practicalpoint of view, substructural annotated logics are in some sense complicated for applications. In fact, we do not know any implementations of substructural annotated logics.

Related to Lewin, Mikenberg and Schwarze's approaches, Rico [142] proposed annotated logics based on *bilattices*, which were extensively studied by some people like Fitting [82, 83]. We will review bilattice logics later.

5.5 Related Systems

Annotated logics can be viewed as a general logical framework for reasoning about incomplete and inconsistent information, in which annotations play a prominent role. However, we can point out that there are such logical frameworks in the literature, which are closely related to annotated logics.

Here, we simply review two frameworks, i.e., *Labelled Deductive Systems* (LDS) due to Gabbay [82] and *General Logics* (GL) due to Slaney [147]. It is noticed that in both frameworks annotation is seriously used.

LDS employ a labelling annotation, appearing not always attached to a wff but in an adjacent column, i.e., $t : A$, where t is a label and A is a formula. A label is formalized algebraically to control the interpretation of a formula.

For example, Gabbay's formulation of the generalized version of *modus ponens* called (MMP) is as follows:

$$\frac{t : A \quad s : A \rightarrow B}{s + t : B}$$

that is, Gabbay's colon : becomes in effect a column spacer.

Gabbay gives a general theory for various implicational logics using the labelled rules, but he does not consider other logical connectives. As it happens, the main illustrative rule he offers is MMP, with coupling relation semi-colon permitted various different rules, such as multiset union.

Gabbay also worked out LDS for other areas like natural language processing and abduction. In this sense, LDS is more general than annotated logics, but Gabbay did not study labelled deductive formulation of paraconsistent logics.

Development of a similar framework was undertaken by Slaney [147]. Approached in a natural deduction way, GL is a relevant generalization of Lemmon's "beginning classical logic", accomplished by distinguishing two ways of compounding labels or assumption codes: by , (extensional) and ; (intensional) coupling. Slaney developed a GL to give a unified account of a fairly wide range of logical systems.

Slaney attempted to elaborate the concept of *bodies of information*. For example, if X and Y are things of the right sort to warrant assertions, then the object $X(Y)$ is the result of taking X as the determination of available inference and applying it to Y.

Another combination of bodies of information is *pooling*, which is very much like forming set union. For instance, $X \cup Y$ is the result of pooling information X with information Y. In addition, Slaney stressed the difference between using information to tell you how the world stands and using it to tell you how you may reason.

For example, a formula $(\sim A \to B) \to (\sim B \to A)$ can be proved in GL as follows:

Accounting columns

Line#	(wff) labels	conclusion of sequent	proof processing record
1.	1	$\sim A \to B$	hypothesis
2.	2	$\sim B$	hypothesis
3.	1; 2	$\sim\sim A$	from 1, 2 by MT
4.	1; 2	A	3, DNE
5.	1	$\sim B \to \sim A$	2, 4, CP
6.	0; 1	$\sim B \to A$	5, SR5
7.	0	$(\sim A \to B) \to (\sim B \to A)$	1, 6, CP

Here, MT denotes *modus tollens*, DNE double negation elimination, CP implication introduction, and SR5 is a structural rule defined below.

Instead of operating simply with wff, objects of investigation now comprise *annotated* wff, syntactically wff coupled with a label. Simple wff are not lost; they are both included, and can be marked out as proved, valid or whatever by distinguishing labels, such as 0. The coupling, that of a many-valued association function, may appear explicitly (e.g. as a colon or like symbol). In short, where B is a wff and c is an annotation, then $c : B$ and $c\ B$, are annotated or labelled wff.

Both wff and annotations conform to (coupled) logics, for annotations so far taking elementary algebraic form. For instance there may be, as in semi-lattice semantics for parts of relevant logic, an operation $^\circ$ on pieces of information which conform to semi-lattice requirements; so the logic of annotation is that of a semi-lattice.

Every algebra can be (variously) represented as a logic; see Routley et al. [144]. On semi-lattice semantics for relevant logic, see Urquhart's contribution in Anderson et al. [35]. In the present main illustration, that borrowed from Slaney, there are two operations on labels; comma and semi-colon. where a and b are labels, so are (a, b) and $(a; b)$.

We promptly dispense with many brackets by what are piquantly called conventions, but amount to a series of rules, many of which can be strengthened to identities. In effect, there is a little logic of brackets, which too may take algebraic form. We have already assumed that we can rewrite complex labels without outer brackets, but in a more complete formalization that would be recorded as rewrite rules, for instance thus:

$(a; b)Ra; b$
$(a, b)Ra, b,$

with R reading: "can be rewritten as".

Such rules do not enable removal or rearrangement of inner brackets. That can be achieved however by identity principles, which algebraically assure replacement.

Thus

$a, b = b, a$
$a, a = a$

whence $a, b/b, a$ (but not conversely; reversible rules are weaker than identities). Also, in the present setting, a comma here works like set union. In different setting of course, a comma could work differently, for instance as a multiset operation or otherwise.

Other rules, which may amount to one-way "identities" allowing internal rearrangement, and which in fact correspond to structural rules of proof-theoretic procedures, are as follows (Slaney's enumeration is adopted):

(SR1) $a, (b, c) \Rightarrow a, b, c$ association to the left
(SR4) $a \Rightarrow a, b$ weakening
(SR5) $0; a = a$ 0-basing

While Slaney takes these as just *abbreviating* structural rules –, here $\varphi \Rightarrow \psi$ abbreviates $\dfrac{-\varphi- : A}{-\psi- : A}$, we are allowing them independence. Naturally, however, we wish to retain, and take advantage of, the tight correspondence.

The same sort of points apply to connections between logical rules in natural deduction form, which allow independence, and these rules in proof-theoretic form, which are not assigned primacy.

While we could reverse Slaney's procedure, and not just for variety, setting out natural deduction rules first, and gathering a Gentzenoid proof-theoretical rule there from. But here we stay with Slaney, presenting the rather obvious proof-theoretic forms first.

Now, we introduce a natural deduction system for the general annotated logic *GAL*. *Proof-theoretic forms* of logical rules in *GAL* are as follows:

Assumption $A : A$

The rule is reexpressed in Slaney, thus $\overline{A{:}A}$ (A). That is, a wff may be introduced on any line of a proof as an assumption. Further logical rules are those for introduction (I) and elimination (E) of connectives from the standard set $\{\&, \vee, \rightarrow, \sim\}$:

$$(\&E)\frac{A\&B}{A} \quad \frac{A\&B}{B} \qquad (\&I)\frac{a : A \quad b : B}{a, b : A\&B}$$

$$(\vee I)\frac{A}{A \vee B} \quad \frac{B}{A \vee B} \qquad (\vee E)\frac{\delta(A) : C \quad \delta(B) : C \quad a : A \vee B}{\delta(a) : C}$$

$$(\rightarrow E(MMP))\frac{a : A \rightarrow B \quad b : A}{a; b : B} \qquad (\rightarrow I(CP))\frac{a; A : B}{a : A \rightarrow B}$$

$$(\sim\sim I)\frac{A}{\sim\sim A} \qquad (\sim\sim E)\frac{\sim\sim A}{A}$$

$$(\sim M(MMT)\frac{a : A \rightarrow B \quad b : \sim B}{a; b : \sim A}$$

Here, $\delta(A)$ means a bunch in which A occurs in some particular place as a subbunch.

Based on semantics-based theory of labels (annotations), GL offers a general framework for interpreting various logics. One of the interesting features of GL is taken as a theory of labels in Routley-Meyer semantics for relevance logics. In contrast, LDS is based on a more flexible theory of labels. In any case, both frameworks try to use label-based logics.

Several ideas in LDS and GL can be incorporated into annotated logics. Although the original version of annotated logics allows an annotation for the atomic formula, the restriction can be eliminated.

In this line, Sylvan and Abe proposed *general annotated logics* in [150]. Rough ideas of general annotated logics are given above. Obviously, general annotated logics can be developed as an interesting extension of annotated logics.

5.6 Systems of Paraconsistent Logics

Because annotated logics are paraconsistent, we need to review systems of paraconsistent logics in the literature. As mentioned in Chap. 1, the study of paraconsistent logics started in the late 1940s and we now have numerous such systems. It is thus impossible to review all approaches to paraconsistent logics. So we here concentrate major systems of paraconsistent logics, i.e. discursive logic, C-systems and relevance logic.

Discursive logic, also known as discussive logic, was proposed by Jaśkowski [95, 96], which is regarded as a non-adjunctive approach. *Adjunction* is a rule of inference of the form: from $\vdash A$ and $\vdash B$ to $\vdash A \wedge B$. Discursive logic can avoid explosion by prohibiting adjunction.

It was a formal system J satisfying the conditions: (a) from two contradictory propositions, it should not be possible to deduce any proposition; (b) most of the classical theses compatible with (a) should be valid; (c) J should have an intuitive interpretation.

Such a calculus has, among others, the following intuitive properties remarked by Jaśkowski himself: suppose that one desires to systematize in only one deductive system all theses defended in a discussion. In general, the participants do not confer the same meaning to some of the symbols.

One would have then as theses of a deductive system that formalize such a discussion, an assertion and its negation, so both are "true" since it has a variation in the sense given to the symbols. It is thus possible to regard discursive logic as one of the so-called *paraconsistent logics*.

Jaśkowski's D_2 contains propositional formulas built from logical symbols of classical logic. In addition, the possibility operator $\Diamond$ in S5 is added. Based on the possibility operator, three discursive logical symbols can be defined as follows:

discursive implication: $A \rightarrow_d B =_{def} \Diamond A \rightarrow B$
discursive conjunction: $A \wedge_d B =_{def} \Diamond A \wedge B$
discursive equivalence: $A \leftrightarrow_d B =_{def} (A \rightarrow_d B) \wedge_d (B \rightarrow_d A)$

Additionally, we can define discursive negation $\neg_d A$ as $A \rightarrow_d$ *false*. Jaśkowski's original formulation of D_2 in [96] used the logical symbols: $\rightarrow_d$, $\leftrightarrow_d$, $\vee$, $\wedge$, $\neg$, and he later defined $\wedge_d$ in [96].

The following axiomatization due to Kotas [104] has the following axioms and the rules of inference.

Axioms

(A1) $\Box(A \rightarrow (\neg A \rightarrow B))$
(A2) $\Box((A \rightarrow B) \rightarrow ((B \rightarrow C) \rightarrow (A \rightarrow C))$
(A3) $\Box((\neg A \rightarrow A) \rightarrow A)$
(A4) $\Box(\Box A \rightarrow A)$
(A5) $\Box(\Box(A \rightarrow B) \rightarrow (\Box A \rightarrow \Box B))$
(A6) $\Box(\neg \Box A \rightarrow \Box \neg \Box A)$

Rules of Inference

(R1) substitution rule
(R2) $\Box A, \Box(A \rightarrow B)/\Box B$
(R3) $\Box A/\Box\Box A$
(R4) $\Box A/A$
(R5) $\neg\Box\neg\Box A/A$

There are other axiomatizations of D_2, but we omit the details here. Discursive logics are considered weak as a paraconsistent logic, but they have some applications, e.g. logics for vagueness.

C-systems are paraconsistent logics due to da Costa which can be a basis for inconsistent but non-trivial theories; see da Costa [61]. The important feature of da Costa systems is to use novel interpretation, which is non-truth-functional, of negation avoiding triviality.

Here, we review C-system C_1 due to da Costa [61]. The language of C_1 is based on the logical symbols: $\wedge$, $\vee$, $\rightarrow$, and $\neg$. $\leftrightarrow$ is defined as usual. In addition, a formula A°, which is read "*A* is well-behaved", is shorthand for $\neg(A \wedge \neg A)$. The basic ideas of C_1 contain the following: (1) most valid formulas in the classical logic hold, (2) the law of non-contradiction $\neg(A \wedge \neg A)$ should not be valid, (3) from two contradictory formulas it should not be possible to deduce any formula.

The Hilbert system of C_1 extends the positive intuitionistic logic with the axioms for negation.

da Costa's C_1

Axioms

(DC1) $A \rightarrow (B \rightarrow A)$
(DC2) $(A \rightarrow B) \rightarrow (A \rightarrow (B \rightarrow C)) \rightarrow (A \rightarrow C))$
(DC3) $(A \wedge B) \rightarrow A$
(DC4) $(A \wedge B) \rightarrow B$
(DC5) $A \rightarrow (B \rightarrow (A \wedge B))$
(DC6) $A \rightarrow (A \vee B)$
(DC7) $B \rightarrow (A \vee B)$
(DC8) $(A \rightarrow C) \rightarrow ((B \rightarrow C) \rightarrow ((A \vee B) \rightarrow C))$
(DC9) $B^\circ \rightarrow ((A \rightarrow B) \rightarrow ((A \rightarrow \neg B) \rightarrow \neg A))$

(DC10) $(A^\circ \wedge B^\circ) \to (A \wedge B)^\circ \wedge (A \vee B)^\circ \wedge (A \to B)^\circ$
(DC11) $A \vee \neg A$
(DC12) $\neg\neg A \to A$

Rules of Inference

(MP) $\vdash A,\ \vdash A \to B \Rightarrow \vdash B$

Here, (DC1)–(DC8) are axioms of the positive intuitionistic logic. (DC9) and (DC10) play a role for the formalization of paraconsistency.

A semantics for C_1 can be given by a two-valued valuation; see da Costa and Alves [61]. We denote by $\mathscr{F}$ the set of formulas of C_1. A valuation is a mapping v from $\mathscr{F}$ to $\{0, 1\}$ satisfying the following:

$$v(A) = 0 \Rightarrow v(\neg A) = 1$$
$$v(\neg\neg A) = 1 \Rightarrow v(A) = 1$$
$$v(B^\circ) = v(A \to B) = v(A \to \neg B) = 1 \Rightarrow v(A) = 0$$
$$v(A \to B) = 1 \Leftrightarrow v(A) = 0 \text{ or } v(B) = 1$$
$$v(A \wedge B) = 1 \Leftrightarrow v(A) = v(B) = 1$$
$$v(A \vee B) = 1 \Leftrightarrow v(A) = 1 \text{ or } v(B) = 1$$
$$v(A^\circ) = v(B^\circ) = 1 \Rightarrow v((A \wedge B)^\circ) = v((A \vee B)^\circ) = v((A \to B)^\circ) = 1$$

Note here that the interpretations of negation and double negation are not given by biconditional. A formula A is *valid*, written $\models A$, if $v(A) = 1$ for every valuation v. Completeness holds for C_1. It can be shown that C_1 is complete for the above semantics.

Da Costa system C_1 can be extended to C_n $(1 \leq n \leq \omega)$. Now, A^1 stands for A° and A^n stands for $A^{n-1} \wedge (A^{(n-1)})^\circ$, $1 \leq n \leq \omega$.

Then, da Costa system C_n $(1 \leq n \leq \omega)$ can be obtained by (DC1)–(DC8), (DC12), (DC13) and the following:

(DC9n) $B^{(n)} \to ((A \to B) \to ((A \to \neg B) \to \neg A))$
(DC10n) $(A^{(n)} \wedge B^{(n)}) \to (A \wedge B)^{(n)} \wedge (A \vee B)^{(n)} \wedge (A \to B)^{(n)}$

Note that the da Costa system C_ω has the axioms (DC1)–(DC8), (DC12) and (DC13). Later, da Costa investigated first-order and higher-order extensions of C-systems.

Relevance logic, also called *relevant logic* is a family of logics based on the notion of relevance in conditionals. Historically, relevance logic was developed to avoid the *paradox of implications*; see Anderson and Belnap [34] and Anderson et al. [35].

Anderson and Belnap formalized a relevant logic R to realize a major motivation, in which they do not admit $A \to (B \to A)$. Later, various relevance logics have been proposed. Note that not all relevance logics are paraconsistent but some are considered important as paraconsistent logics.

Routley and Meyer proposed a basic relevant logic B, which is a minimal system having the so-called *Routley-Meyer semantics*. Thus, B is an important system and we review it below; see Routley et al. [144].

The language of B contains logical symbols: $\sim$, &, $\vee$ and $\to$ (relevant implication). A Hilbert system for B is as follows:

Relevant Logic B

Axioms

(BA1) $A \to A$
(BA2) $(A\&B) \to A$
(BA3) $(A\&B) \to B$
(BA4) $((A \to B)\&(A \to C)) \to (A \to (B\&C))$
(BA5) $A \to (A \vee B)$
(BA6) $B \to (A \vee B)$
(BA7) $(A \to C)\&(B \to C)) \to ((A \vee B) \to C)$
(BA8) $(A\&(B \vee C)) \to (A\&B) \vee C)$
(BA9) $\sim\sim A \to A$

Rules of Inference

(BR1) $\vdash A, \vdash A \to B \Rightarrow \vdash B$
(BR2) $\vdash A, \vdash B \Rightarrow \vdash A\&B$
(BR3) $\vdash A \to B, \vdash C \to D \Rightarrow \vdash (B \to C) \to (A \to D)$
(BR4) $\vdash A \to \sim B \Rightarrow \vdash B \to \sim A$

A Hilbert system for Anderson and Belnap's R is as follows:

Relevance Logic R

Axioms

(RA1) $A \to A$
(RA2) $(A \to B) \to ((C \to A) \to C \to B))$
(RA3) $(A \to (A \to B) \to (A \to B)$
(RA4) $(A \to (B \to C)) \to (B \to (A \to C)$
(RA5) $(A\&B) \to A$
(RA6) $(A\&B) \to B$
(RA7) $((A \to B)\&(A \to C)) \to (A \to (B\&C))$
(RA8) $A \to (A \vee B)$
(RA9) $B \to (A \vee B)$
(RA10) $((A \to C)\&(B \vee C)) \to ((A \vee B) \to C))$
(RA11) $(A\&(B \vee C)) \to ((A\&B) \vee C)$
(RA12) $(A \to\sim A) \to\sim A$
(RA13) $(A \to\sim B)) \to (B \to\sim A)$
(RA14) $\sim\sim A \to A$

Rules of Inference

(RR1) $\vdash A, \vdash A \to B \Rightarrow \vdash B$
(RR2) $\vdash A, \vdash B \Rightarrow \vdash A\&B$

Routley et al. considered some axioms of R are too strong and formalized rules instead of axioms. Notice that B is a paraconsistent but R is not.

Next, we give a Routley-Meyer semantics for B. A model structure is a tuple $\mathcal{M} = \langle K, N, R, *, v \rangle$, where K is a non-empty set of worlds, $N \subseteq K$, $R \subseteq K^3$ is a ternary relation on K, $*$ is a unary operation on K, and v is a valuation function from a set of worlds and a set of propositional variables $\mathcal{P}$ to $\{0, 1\}$.

There are some restrictions on . v satisfies the condition that $a \leq b$ and $v(a, p)$ imply $v(b, p) = 1$ for any $a, b \in K$ and any $p \in \mathcal{P}$. $a \leq b$ is a pre-order relation defined by $\exists x(x \in N$ and $Rxab)$. The operation $*$ satisfies the condition $a^{**} = a$.

For any propositional variable p, the truth condition $\models$ is defined: $a \models p$ iff $v(a, p) = 1$. Here, $a \models p$ reads "p is true at a". $\models$ can be extended for any formulas in the following way:

$$a \models \sim A \Leftrightarrow a^* \not\models A$$
$$a \models A \& B \Leftrightarrow a \models A \text{ and } a \models B$$
$$a \models A \vee B \Leftrightarrow a \models A \text{ or } a \models B$$
$$a \models A \rightarrow B \Leftrightarrow \forall bc \in K(Rabc \text{ and } b \models A \Rightarrow c \models B)$$

A formula A is *true* at a in $\mathscr{M}$ iff $a \models A$. A is *valid*, written $\models A$, iff A is true on all members of N in all model structures.

Routley et al. provides the completeness theorem for B with respect to the above semantics using canonical models; see [144].

A model structure for R needs the following conditions.

$$R0aa$$
$$Rabc \Rightarrow Rbac$$
$$R^2(ab)cd \Rightarrow R^2a(bc)d\ Raaa$$
$$a^{**} = a$$
$$Rabc \Rightarrow Rac^*b^*$$
$$Rabc \Rightarrow a' \leq a \Rightarrow Ra'bc$$

where R^2abcd is shorthand for $\exists x(Raxd$ and $Rxcd)$. The completeness theorem for the Routley-Meyer semantics can be proved for R; see [34, 35].

The reader is advised to consult Anderson and Belnap [34], Anderson et al. [35], and Routley et al. [144] for details. A more concise survey on the subject may be found in Dunn [77].

Belnap proposed a famous *four-valued logic* in Belnap [49, 50], which is closely related to relevant logic and paraconsistent logic. Belnap's four-valued logic aims to formalize the internal states of a computer.

There are four states, i.e. (*T*), (*F*), (*None*) and (*Both*), to recognize an input in a computer. Based on these states, a computer can compute the following suitable outputs.

(*T*) a proposition is true.

(*F*) a proposition is false.

(*N*) a proposition is neither true nor false.

(*B*) a proposition is both true and false.

Here, (*N*) and (*B*) abbreviate (*None*) and (*Both*), respectively. From the above, (*N*) corresponds to incompleteness and (*B*) inconsistency. Four-valued logic can be thus seen as a natural extension of three-valued logic. In fact, Belnap's four-valued logic can model both incomplete information (*N*) and inconsistent information (*B*).

Belnap proposed two four-valued logics **A4** and **L4**. The former can cope only with atomic formulas, whereas the latter can handle compound formulas. **A4** is based on the *approximation lattice*, which is shown in Fig. 5.1. This is the same as Fig. 2.1.

Here, B is the least upper bound and N is the greatest lower bound with respect to the ordering $\leq$. Observe that in the lattice *FOUR* in Fig. 5.1, we used $t, f, \bot, \top$ instead of T, F, N, B, respectively.

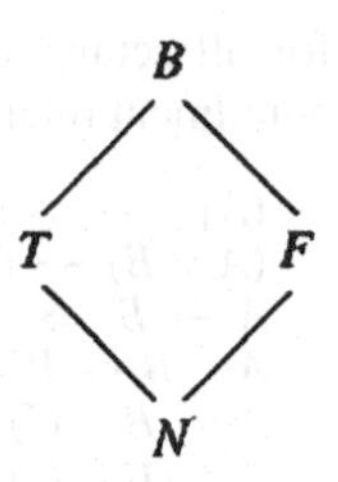

Fig. 5.1 Approximation lattice **L4**

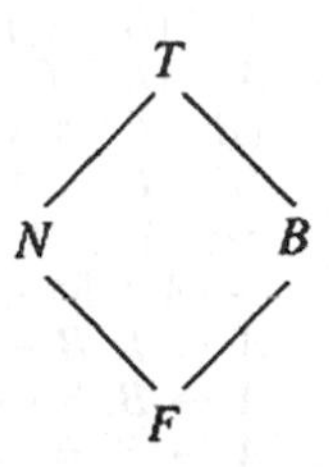

Fig. 5.2 Logical lattice **L4**

The logic **L4** has logical symbols; $\sim, \wedge, \vee$. Its truth-values is $\mathbf{4} = \{T, F, N, B\}$ with a different ordering. The lattice **L4** is shown in Fig. 5.2.

One of the features of **L4** is the monotonicity of logical symbols. Let f be a logical operation. It is said that f is monotonic iff $a \subseteq b \Rightarrow f(a) \subseteq f(b)$. To guarantee the monotonicity of conjunction and disjunction, they must satisfy the following:

$a \wedge b = a \Leftrightarrow a \vee b = b$
$a \wedge b = b \Leftrightarrow a \vee b = a$

Logical symbols in **L4** obey th truth-value tables in Table 5.1.

Belnap gave a semantics for the language with the above logical symbols. A *setup* is a mapping a set of atomic formulas *Atom* to the set **4**. Then, formulas of **L4** are defined as follows:

$s(A \wedge B) = s(A) \wedge s(B)$
$s(A \vee B) = s(A) \vee s(B)$
$s(\sim A) = \sim s(A)$

Further, Belnap defined an entailment relation $\rightarrow$ as follows:

$A \rightarrow B \Leftrightarrow s(A) \leq s(B)$

Table 5.1 Truth-value tables of **L4**

	N	F	T	B
$\sim$	B	T	F	N

$\wedge$	N	F	T	B
N	N	F	N	F
F	F	F	F	F
T	N	F	T	B
B	F	F	B	B

$\vee$	N	F	T	B
N	N	N	T	T
F	N	F	T	B
T	T	T	T	T
B	T	B	T	B

for all setups s. Note that $\to$ is not a logical connective for implication but an entailment relation. The entailment relation $\to$ can be axiomatized as follows:

$$(A_1 \wedge \cdots \wedge A_m) \to (B_1 \vee \cdots \vee B_n) \ (A_i \text{ shares some } B_j)$$
$$(A \vee B) \to C \leftrightarrow (A \to C) \text{ and } (B \to C)$$
$$A \to B \Leftrightarrow \sim B \to \sim A$$
$$A \vee B \leftrightarrow B \vee A, \ A \wedge B \leftrightarrow B \wedge A$$
$$A \vee (B \vee C) \leftrightarrow (A \vee B) \vee C$$
$$A \wedge (B \wedge C) \leftrightarrow (A \wedge B) \wedge C$$
$$A \wedge (B \vee C) \leftrightarrow (A \wedge B) \vee (A \wedge C)$$
$$A \vee (B \wedge C) \leftrightarrow (A \vee B) \wedge (A \vee C)$$
$$(B \vee C) \wedge A \leftrightarrow (B \wedge A) \vee (C \wedge A)$$
$$(B \wedge C) \vee A \leftrightarrow (B \vee A) \wedge (C \vee A)$$
$$\sim\sim A \leftrightarrow A$$
$$\sim (A \wedge B) \leftrightarrow \sim A \vee \sim B, \ \sim (A \vee B) \leftrightarrow \sim A \wedge \sim B$$
$$A \to B, B \to C \Leftrightarrow A \to C$$
$$A \leftrightarrow B, B \leftrightarrow C \Leftrightarrow A \leftrightarrow C$$
$$A \to B \Leftrightarrow A \leftrightarrow (A \wedge B) \Leftrightarrow (A \vee B) \leftrightarrow B$$

Note here that $(A \wedge \sim A) \to B$ and $A \to (B \vee \sim B)$ cannot be derived in this axiomatization. It can be shown that the logic given above is shown to be equivalent to the system of *tautological entailment*; see [34, 35].

An alternative semantics for tautological entailment based on the notion of fact was worked out by van Fraassen [155]. Belnap's **A4** is used as one of the lattice of truth-values as *FOUR*. In this regard, Belnap's four-valued logic is considered as the important background on annotated logics.

Although the above three logics are famous approaches to paraconsistent logics, there is a rich literature on paraconsistent logics. Arruda [39] reviewed a survey on paraconsistent logics, and Priest et al. [136] contains interesting papers on paraconsistent logics in the 1980s. For a recent survey, we refer Priest [138]. We can also find a Handbook surveying various subjects related to paraconistency by Beziau et al. [52].

In 1997, The First World Congress on Paraconsistency (WCP'1997) was held at the University of Ghent, Belgium; see Batens et al. [46]. The Second World Congress on Paraconsistency (WCP'200) was held at Juquehy-Sao Paulo, Brazil; see Carnielli et al. [55].

In the 1990s paraconsistent logics became one of the major topics in logic in connection with other areas, in particular, computer science. Below we review some of those systems of paraconsistent logics.

As mentioned in Chap. 1, the modern history of paraconsistent logic started with Vasil'ev's *imaginary logic*. In 1910, Vasil'ev proposed an extension of Aristotle's syllogistic allowing the statement of the form S is both P and not-P; see Vasil'ev [156].

Thus, imaginary logic can be viewed as a paraconsistent logic. Unfortunately, little work has been done on focusing on its formalization from the viewpoint of modern logic. A survey of imaginary logic can be found in Arruda [38].

In 1954, Asenjo developed a calculus of antinomies in his dissertation; see Asenjo [40]. Asenjo's work was published before da Costa's work, but it seems that Asenjo's approach has been neglected. Asenjo's idea is to interpret the truth-value of *antinomy* as both true and false using Kleene's strong three-valued logic.

His proposed calculus is non-trivially inconsistent propositional logic, whose axiomatization can be obtained from Kleene's [103] axiomatization of classical propositional logic by deleting the axiom $(A \rightarrow B) \rightarrow ((A \rightarrow \neg B) \rightarrow \neg A)$.

In constructivism, an idea of constructing paraconsistent logics may be found. In 1949, Nelson [132] proposed a *constructive logic with strong negation* as an alternative to intuitionistic logic, in which *strong negation* (or constructible negation) is introduced to improve some weaknesses of intuitionistic negation.[2]

Constructive logic N extends positive intuitionistic logic Int^+ with the following axioms for *strong negation* $\sim$:

(N1) $(A \wedge \sim A) \rightarrow B$
(N2) $\sim\sim A \leftrightarrow A$
(N3) $\sim (A \rightarrow B) \leftrightarrow (A \wedge \sim B)$
(N4) $\sim (A \wedge B) \leftrightarrow (\sim A \vee \sim B)$
(N5) $\sim (A \vee B) \leftrightarrow (\sim A \wedge \sim B)$

In N, intuitionistic negation $\neg$ can be defined as $\neg A \leftrightarrow A \rightarrow (B \wedge \sim B)$. If we delete (N1) from N, we can obtain a paraconsistent constructive logic N^- of Almukdad and Nelson [32]. Akama [16–21] extensively studied Nelson's constructive logics with strong negation; also see Wansing [157].

In 1959, Nelson [133] developed a constructive logic S which lacks contraction $(A \rightarrow (A \rightarrow B)) \rightarrow (A \rightarrow B)$ and discussed its aspects as a paraconsistent logic. Akama [20] gave a detailed presentation of Nelson's paraconsistent constructive logics. Akama et al. [30] proposed a constructive discursive logic based on Nelson's constructive logic.

In 1979, Priest [137] proposed a *logic of paradox*, denoted *LP*, to deal with the semantic paradox. The logic is of special importance to the area of paraconsistent logics. *LP* can be semantically defined by Kleene's strong three-valued logic whose truth-value tables are as Table 5.2.

Here, T and F denote truth and falsity, and the third truth-value I reads "undefined"; see Kleene [103].

Łukasiewicz's three-valued logic is interpreted by the above truth-value tables of Kleene's three-valued logic except for implication. Let $\rightarrow_L$ be the implication in Łukasiewicz's three-valued logic. Then, the truth-value tables are described as Table 5.3.

Here, the third truth-value reads "possible"; see Łukasiewicz [116]. Kleene's three-valued logic was used as a basis for reasoning about incomplete information in computer science.

Priest re-interpreted the truth-value tables of Kleene's strong three-valued logic, namely read the third-truth value as both true and false (B) rather than neither true nor false (I), and assumed that (T) and (B) are designated values. The idea has already been considered in Asenjo [40] and Belnap [49, 50].

Consequently, ECQ: $A, \sim A \models B$ is invalid. Thus, *LP* can be seen as a paraconsistent logic. Unfortunately, (material) implication in *LP* does not satisfy *modus*

[2] Here, Nelson's strong negation should be conceptually distinguished from the one in annotated logics.

Table 5.2 Truth-value tables of Kleene's strong three-valued logic

A	$\neg A$
T	F
I	I
F	T

A	B	$A \wedge B$	$A \vee B$	$A \rightarrow_K B$
T	T	T	T	T
T	F	F	T	F
T	I	I	T	I
F	T	F	F	T
F	F	F	F	T
F	I	F	I	T
I	T	I	T	T
I	F	F	I	I
I	I	I	I	I

Table 5.3 Truth-value tables of Łukasiewicz's three-valued logic

A	$\sim A$
T	F
I	I
F	T

A	B	$A \wedge B$	$A \vee B$	$A \rightarrow_L B$
T	T	T	T	T
T	F	F	T	F
T	I	I	T	I
F	T	F	F	T
F	F	F	F	T
F	I	F	I	T
I	T	I	T	T
I	F	F	I	I
I	I	I	I	T

ponens. It is, however, possible to introduce relevant implications as real implication into *LP*.

Priest developed a semantics for *LP* by means of a truth-value assignment relation rather than a truth-value assignment function. Let $\mathscr{P}$ be the set of propositional variables. Then, an evaluation η is a subset of $\mathscr{P} \times \{0, 1\}$. A proposition may only relate to 1 (true), it may only relate to 0 (false), it may relate to both 1 and 0 or it may relate to neither 1 nor 0. The evaluation is extended to a relation for all formulas as follows:

$\neg A \eta 1$ iff $A \eta 0$
$\neg A \eta 0$ iff $A \eta 1$
$A \wedge B \eta 1$ iff $A \eta 1$ and $B \eta 1$

$A \wedge B\eta 0$ iff $A\eta 0$ or $B\eta 0$
$A \vee B\eta 1$ iff $A\eta 1$ or $B\eta 1$
$A \vee B\eta 0$ iff $A\eta 0$ and $B\eta 0$

If we define validity in terms of truth preservation under all relational evaluations, then we obtain *first-degree entailment* which is a fragment of relevance logics.

Using *LP*, Priest advanced his research program to tackle various philosophical and logical issues; see Priest [138, 139] for details. For instance, in *LP*, the liar sentence can be interpreted as both true and false. It is also observed that Priest promoted the philosohical view called *dialetheism* which claims that there are true contradictions. In fact, dialetheism has been extensively discussed by many people.

Since the beginning of the 1990s, Batens developed the so-called *adaptative logics* in Batens [44, 45]. These logics are considered as improvements of *dynamic dialectical logics* investigated in Batens [43]. *Inconsistency-adaptive logics* as developed by Batens [44] can serve as foundations for paraconsistent and non-monotonic logics.

Adaptive logics formalized classical logic as "dynamic logic".[3] A logic is *adaptive* iff it adapts itself to the specific premises to which it is applied. In this sense, adaptive logics can model the dynamics of human reasoning. There are two sorts of dynamics, i.e., *external dynamics* and *internal dynamics*.

The external dynamics is stated as follows. If new premises become available, then consequences derived from the earlier premise set may be withdrawn. In other words, the external dynamics results from the *non-monotonic* character of the consequence relations.

Let $\vdash$ be a consequence relation, Γ, Δ be sets of formulas, and A be a formula. Then, the external dynamics is formally presented as: $\Gamma \vdash A$ but $\Gamma \cup \Delta \nvdash A$ for some Γ, Δ and A. In fact, the external dynamics is closely related to the notion of *non-monotonic reasoning* in AI.

The internal dynamics is very different from the external one. Even if the premise set is constant, certain formulas are considered as derived at some stage of the reasoning process, but are considered as not derived at a later stage. For any consequence relation, insight in the premises is gained by deriving consequencces from them.

In the absence of a positive test, this results in the internal dynamics. Namely, in the internal dynamics, reasoning has to adapt itself by withdrawing an application of the previously used inference rule, if we infer a contradiction at a later stage. Adaptive logics are logics based on the internal dynamics.

An Adaptive Logic *AL* can be characterized as a triple:

(i) A *lower limit logic* (LLL)
(ii) A set of *abnormalities*
(iii) An *adaptive strategy*

The lower limit logic *LLL* is any monotonic logic, e.g., classical logic, which is the stable part of the adaptive logic. Thus, *LLL* is not subject to adaptation. The set of

[3] Dynamic logic here is not the family of logics with the same name studied in computer science.

abnormalities Ω comprises the formulas that are presupposed to be false, unless and until proven otherwise.

In many adaptive logics, Ω is the set of formulas of the form $A \wedge \sim A$. An adaptive strategy specifies a strategy of the applications of inference rules based on the set of abnormalities.

If the lower limit logic *LLL* is extended with the requirement that no abnormality is logically possible, one obtains a monotonic logic, which is called the *upper limit logic ULL*. Semantically, an adequate semantics for the upper limit logic can be obtained by selecting those lower limit logic models that verify no abnormality. The name "abnormality" refers to the upper limit logic. *ULL* requires premise sets to be normal, and 'explodes' abnormal premise sets (assigns them the trivial consequence set).

If the lower limit logic is classical logic *CL* and the set of abnormalities comprises formulas of the form $\exists A \wedge \exists \sim A$, then the upper limit logic obtained by adding to *CL* the axioms $\exists A \rightarrow \forall A$. If, as is the case for many inconsistency-adaptive logics, the lower limit logic is a paraconsistent logic *PL* which contains *CL*, and the set of abnormalities comprises the formulas of the form $\exists(A \wedge \sim A)$, then the upper limit logic is *CL*. The adaptive logics interpret the set of premises 'as much as possible' in agreement with the upper limit logic; it avoids abnormalities 'in as far as the premises permit'.

Adaptive logics provide a new way of thinking of the formalization of paraconsistent logics in view of the dynamics of reasoning. Although inconsistency-adaptive logic is paraconsistent logic, applications of adaptive logics are not limited to paraconsistency. From a formal point of view, we can count adaptive logics as promising paraconsistent logics.

However, for applications, we may face several obstacles in automating reasoning in adaptive logics in that proofs in adaptive logics are dynamic with a certain adaptive strategy. Thus, the implementation is not easy, and we have to choose an appropriate adaptive strategy depending on applications.

Carnelli proposed the *Logics of Formal Inconsistency* (LFI), which are logical systems that treat consistency and inconsistency as mathematical objects; see Carnelli et al. [56].

One of the distinguishing features of these logics is that they can internalize the notions of consistency and inconsistency at the object-level. And many paraconsistent logics including da Costa's C-systems can be interpreted as the subclass of LFIs. Therefore, we can regard LFIs as a general framework for paraconsistent logics.

A Logic of Formal Inconsistency, which extends classical logic C with the consistency operator $\circ$, is defined as any explosive paraconsistent logic, namely iff the classical consequence relation $\vdash$ satisfies the following two conditions:

(a) $\exists \Gamma \exists A \exists B(\Gamma, A, \neg A \nvdash B)$
(b) $\forall \Gamma \forall A \forall B)(\Gamma, \circ A, A, \neg A \vdash B)$.

Here, Γ denotes a set of formulas and A, B are formulas. With the help of $\circ$, we can express both consistency and inconsistency in the object-language. Therefore, LFIs are general enough to classify many paraconsistent logics.

For example, da Costa's C_1 is shown to be an LFI. For every formula A, let $\circ A$ be an abbreviation of the formula $\neg(A \wedge \neg A)$. Then, the logic C_1 is an LFI such that $\circ(p) = \{\circ p\} = \{\neg\neg(p \wedge \neg p\}$ whose axiomatization as an LFI contains the positive fragment of classical logic with the axiom $\neg\neg A \rightarrow A$, and some axioms for $\circ$.

(bc1) $\circ A \rightarrow (A \rightarrow (\neg A \rightarrow B))$
(ca1) $(\circ A \wedge \circ B) \rightarrow \circ(A \wedge B)$
(ca2) $(\circ A \wedge \circ B) \rightarrow \circ(A \vee B)$
(ca3) $(\circ A \wedge \circ B) \rightarrow \circ(A \rightarrow B)$

In addition, we can define classical negation $\sim$ by $\sim A =_{def} \neg A \wedge \circ A$. If needed, the inconsistency operator $\bullet$ is introduced by definition: $\bullet A =_{def} \neg \circ A$.

Carnielli et al. [56] showed classifications of existing logical systems. For example, classical logic is not an LFI, and Jáskowski's D_2 is an LFI. They also introduced a basic system of LFI, called LFI1 with a semantics and axiomatization.

We can thus see that the Logics of Formal Inconsistency are very interesting from a logical point of view. In addition, there are tableau systems for LFIs; see Carnielli and Marcos [57], and they can be properly applied to various areas including computer science.

The above mentioned logics have been worked as a paraconsistent logic. But there are other logics which share the features of paraconsistent logics. The two notable examples are *possibilistic logic* and logics based on *bilattices*. In fact, these logics can properly deal both with incopmleteness and inconsistency of information. As we reviewed the former in Sect. 5.2, we here look at the latter.

A *bilattice* was originally introduced by Ginsberg [86, 87] for the foundations of reasoning in AI, which has two kinds of orderings, i.e., truth ordering and knowledge ordering. Later, it was extensively studied by Fitting in the context of logic programming in [82] and of theory of truth in [83]. In fact, bilattice-based logics can handle both incomplete and inconsistent information.

A *pre-bilattice* is a structure $\mathscr{B} = \langle B, \leq_t, \leq_k \rangle$, where B denotes a non-empty set and $\leq_t$ and $\leq_k$ are partial orderings on B. The ordering $\leq_k$ is thought of as ranking "degree of information (or knowledge)". The bottom in $\leq_k$ is denoted by $\bot$ and the top by $\top$. If $x <_k y$, y gives us at least as much information as x (and possibly more).

The ordering $\leq_t$ is an ordering on the "degree of truth". The bottom in $\leq_t$ is denoted by *false* and the top by *true*. A bilattice can be obtained by adding certain assumptions for connections for two orderings.

One of the most well-known bilattices is the bilattice *FOUR* as depicted as Fig. 5.3. The billatice *FOUR* can be interpreterd a combination of Belap's lattices **A4** and **L4** as is clear from Fig. 5.3.

The bilattice *FOUR* can be seen as Belnap's lattice *FOUR* with two kinds of orderings. Thus, we can think of the left-right direction as characterizing the ordering $\leq_t$: a move to the right is an increase in truth. The meet operation $\wedge$ for $\leq_t$ is then characterized by: $x \wedge y$ is rightmost thing that is of left both x and y. The join operation $\vee$ is dual to this. In a similar way, the up-down direction characterizes $\leq_k$: a move up is an increase in information. $x \otimes y$ is the uppermost thing below both x and y, and $\oplus$ is its dual.

Fig. 5.3 The bilattice *FOUR*

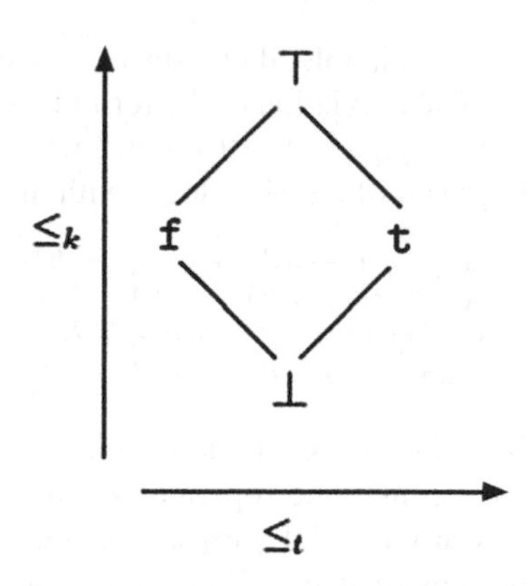

Fitting [82] gave a semantics for logic programming using bilattices. Kifer and Subrahmanian [100] interpreted Fitting's semantics within generalized annotated logics *GAL*. Fitting [83] tried to generalize Kripke's [107] theory of truth, which is based on Kleene's strong three-valued logic, in a four-valued setting based on the bilattice *FOUR*.

A billatice has a negation operation $\neg$ if there is a mapping $\neg$ that reverse $\leq_t$, leaves unchanged $\leq_k$ and $\neg\neg x = x$. Likewise a bilattice has a *conflation* if there is a mapping $-$ that reverse $\leq_k$, leaves unchanged $\leq_t$. and $--x = x$. If a bilattice has both operations, they *commute* if $-\neg x = \neg - x$ for all x.

In the billatice *FOUR*, there is a negation operator under which $\neg t = f$, $\neg f = t$, and $\bot$ and $\top$ are left unchanged. There is also a conflation under which $-\bot = \top$, $-\top = \bot$ and t and f are left unchanged. And negation and conflation commute. In any bilattice, if a negation or conflation exists then the extreme elements $\bot, \top, f$ and t will behave as in *FOUR*.

Bilattice logics are theoretically elegant in that we can obtain several algebraic constructions, and are also suitable for reasoning about incomplete and inconsistent information. Arieli and Avron [36, 37] studied reasoning with bilattices. Thus, bilattice logics have many applications in AI as well as philosophy. In this sense, billatice logics are a rival to annotated logics. As mentioned above, we can also unify annotated logics and billatice logics; see Rico [142].

Chapter 6
Applications

Abstract This chapter discusses applications of annotated logics for various areas. After reviewing paraconsistent logic programming and generalized annotated logic programming, we survey promising applications to knowledge representation, neural computing, automation and robotics.

6.1 Paraconsistent Logic Programming

The origin of annotated logics is closely tied with *logic programming*. In 1974, Kowalski [105] proposed to interpret a Horn clause subset of first-order predicate logic as a programming language. The computational procedure is given by *resolution* due to Robinson [143]. Later, the logic programming language *Prolog* has been used for practical applications.

Although Prolog was useful to deal with many problems, there are some limitations in first-order logic programming, and a lot of work has been done to extend Prolog. One such limitation is obviously the treatment of incomplete and inconsistent information, and the solution is to develop *paraconsistent logic programming*.

From this observation, in 1997, Subrahmanian [149] proposed annotated logics for qualitative logic programming. Later, Blair and Subrahmanian [53] provided the foundations for paraconsistent logic programming. A similar approach to logic programming may be found in van Emden [153]. As we pointed out before, in the 1990s, foundational studies of annotated logics began.

Here, we review paraconsistent logic programming. The motivation of paraconsistent logic programming is clear. Blair and Subrahmanian [53] said:

> However, if logic programming is to be a *pragmatic* tool for the development of knowledge bases, it must have some means for dealing with inconsistent knowledge. Take for example an expert system developed by a team of logic programmers. Each programmer might have acquired information from various domain experts. It is very common for experts to disagree (often strongly). Thus, the knowledge base so developed might contain inconsistent information.

The quotation nicely explains the need for paraconsistent logic programming in AI applications. And this is common to any paraconsistent logics.

J.M. Abe et al., *Introduction to Annotated Logics*,
Intelligent Systems Reference Library 88, DOI 10.1007/978-3-319-17912-4_6

Now, we formally introduce paraconsistent logic programming following [53]. The set $\mathcal{T} = \{\bot, \mathbf{t}, \mathbf{f}, \top\}$ is the set of *truth-values* of a four-valued logic. The logic was proposed by Belnap [49, 50] to model states of a computer. Truth-values $\bot, \mathbf{t}, \mathbf{f}, \top$ respectively stand for "undefined", true, false, and "over-defined" in the intuition of two-valued logic.

Definition 6.1 $\forall x, y \in \mathcal{T}, x \leq y \Leftrightarrow x = y \vee x = \bot \vee y = \top$.

The set $\mathcal{T}$ is a complete lattice under the ordering $\leq$. A *literal* is either an atomic formula or its negation. Following the notation in logic programming, implication is denoted by $\Leftarrow$.

Definition 6.2 If A is a literal, then $A : \mu$ is called an *annotated literal*, where $\mu \in \mathcal{T}$. μ is called the *annotation* of A. If μ is one of $\mathbf{t}, \mathbf{f}$, then $A : \mu$ is called a *well-annotated* literal, and μ is called a w-*annotation*.

The annotated atom $A : \mathbf{t}$ may be read as a rough approximation to "A is *known* to be true" and, similarly $A : \mathbf{f}$ may be as regarded as saying "A is *known* to be false".

Definition 6.3 (*Formulas*)

1. Any annotated atom is a *formula*. Atoms are the usual atomic formulas of first-order logic.
2. $A : \mu$ is an annotated atom, then $\neg A : \mu$ is a formula.
3. If F_1, F_2 are formulas, then $F_1 \& F_2$ (conjunction), $F_1 \vee F_2$ (disjunction), $F_1 \Leftarrow F_2$ (implication), $F_1 \Leftrightarrow F_2$ (equivalence) are formulas.
4. If F is a formula and x any variable symbol, then $(\forall x)F$ and $(\exists x)F$ are formulas.

The equivalence is defined as usual. The following definition defines a generalization of Horn clause.

Definition 6.4 (*Generalized Horn clause*) If $A_0, \ldots, A_n$ are literals, and if $\mu_0, \ldots, \mu_n$ are w-annotations, then

$$A_0 : \mu_0 \Leftarrow A_1 : \mu_1 \& \ldots \& A_n : \mu_n$$

is called a *generalized Horn clause* (or gh-clause for short). $A_0 : \mu_0$ is called the *head* of the above gh-clause, while $A_1 : \mu_1 \& \ldots \& A_n : \mu_n$ is called the *body* of the above gh-clause.

The notion of a substitution is as usual. Applying a substitution θ to an annotated literal $a : \mu$ results in the annotated literal $A\theta : \mu$, and it can be extended to a conjunction of annotated literals and to gh-clauses.

Definition 6.5 If $A : \mu$ and $B : \rho$ are annotated literals, then $A : \mu$ and $B : \rho$ are said to be *unifiable* (with mgu θ) iff A and B are unifiable (with mgu θ). Note that μ need not equal ρ as we are not defining the result of the unification yet.

Definition 6.6 A *generalized Horn program* (GHP) is a finite set of gh-clauses.

Blair and Subrahmanian considered a model-theoretic semantics for GHP based on "Herbrand-like" interpretations, a fixpoint semantics and an operational semantics. These three are the semantics often discussed in the area of logic programming. For classical Horn clause logic programming, van Emden and Kowalski first established these three semantics in [154].

Let I be an interpretation $I : B_L \rightarrow \mathcal{T}$, where B_L is the Herbrand base under consideration.

Definition 6.7 The *negation* of an annotation is defined as: $\neg(\mathbf{t}) = \mathbf{f}, \neg(\mathbf{f}) = \mathbf{t}, \neg(\bot) = \bot, \neg(\top) = \top$.

Definition 6.8 A formula is said to be *closed* if it contains no occurrence of a free variable.

We denote the universal closure and existential closure of the formula F by $(\forall)F$ and $(\exists)F$, respectively.

It is observed that $\Leftarrow$ has a non-classical interpretation in that we do not treat it as classical material implication. In particular, the equivalence of $A \vee \neg B$ and $A \Leftarrow B$ does not hold.

Definition 6.9 (*Satisfaction*) We write $I \models F$ to say that I satisfies F. An interpretation I

1. satisfies the formulas F iff I satisfies each of its closed instances, i.e., for each variable symbol x occurring free in F, and each variable free term t, $F(t/x)$ is satisfied by I (here $F(t/x)$ denotes the replacement of all free occurrences of x in F by t),
2. satisfies the closed annotation atom $A : \mu$ iff $I(A) \geq \mu$,
3. satisfies the closed annotation literal $\neg A : \mu$ iff it satisfies $A : \neg\mu$,
4. satisfies the closed formula $(\exists x)F$ iff for some variable free term t, $I \models F(t/x)$,
5. satisfies the closed formula $(\forall x)F$ iff for every variable free term t, $I \models F(t/x)$,
6. satisfies the closed formula $F_1 \Leftarrow F_2$ iff $I \not\models F_2$ or $I \models F_1$,
7. satisfies the closed formula $F_1 \& \ldots \& F_n$ iff $I \models F_i$ for all $i = 1, \ldots, n$,
8. satisfies the closed formula $F_1 \vee \ldots \vee F_n$ iff $I \models F_i$ for some $1 \leq i \leq n$,
9. satisfies the closed formula $F \Leftrightarrow G$ iff $I \models F \Leftarrow G$ and $I \models G \Leftarrow F$.

Definition 6.10 An interpretation I satisfies a generalized Horn program G if it satisfies every gh-clause $C \in G$.

The ordering $\leq$ on truth-values is extended for interpretations. Given a GHP G and Herbrand interpretations I_1, I_2, we say

$I_1 \leq I_2$ iff $(\forall A \in B_G) I_1(A) \leq I_2(A)$

where B_G is the Herbrand base of G. Note that the set of interpretations (over the set $\mathcal{T}$ of truth-values) is a complete lattice under the ordering $\leq$.

Definition 6.11 If C is a gh-clause in G, then the result of replacing all negated literals $\neg A : \mu$ in C by $A : \neg\mu$ is called the *positive counterpart* C^{pos} of C. The GHP G^{pos} obtained by replacing each gh-clause C in the GHP G by C^{pos} is called the *positive counterpart* of G and is denoted by G^{pos}.

Lemma 6.1 *The following hold:*

$I \models (\exists)\neg A : \mu$ *iff* $I \models (\exists) A : \neg\mu$,
$I \models (\forall)\neg A : \mu$ *iff* $I \models (\forall) A : \neg\mu$.

Proof Let $\overrightarrow{t}$ is a tuple of free terms and $\overrightarrow{x}$ is a tuple of all the variable symbols occurring in A.

$$\begin{aligned} I \models (\exists)\neg A : \mu \text{ iff } & I \models \neg A(\overrightarrow{t}/\overrightarrow{x}) : \mu \text{ for some tuple } \overrightarrow{t} \text{ of variable} - \text{free terms} \\ \text{iff } & I \models A(\overrightarrow{t}/\overrightarrow{x}) : \neg\mu \text{ for some tuple } \overrightarrow{t} \text{ of variable} - \text{free terms} \\ \text{iff } & I \models (\exists) A : \neg\mu \end{aligned}$$

The proof for the universally quantified case is similar.

Theorem 6.1 *I is a model of G iff I is a model of* G^{pos}.

From Theorem 6.1, we can dispense with negated literals in GHPs.

Example 6.1 For example, consider the following GHP G:

$p(a) : \mathbf{t} \Leftarrow, \quad p(a) : \mathbf{f} \Leftarrow$

G has exactly one model, namely the interpretation that assigns the truth-value $\top$ to $p(a)$. This reveals that G contains contradictory information.

Example 6.2 Consider GHP G:

$p(a) : \mathbf{t} \Leftarrow p(b) : \mathbf{f}$,
$p(b) : \mathbf{t} \Leftarrow p(a) : \mathbf{f}$

G has several models as follows:

$$\begin{aligned}
&I_1 : I_1(p(a)) = \mathbf{t}, \quad I_1(p(b)) = \mathbf{t} \\
&I_2 : I_2(p(a)) = \mathbf{t}, \quad I_2(p(b)) = \mathbf{f} \\
&I_3 : I_3(p(a)) = \mathbf{f}, \quad I_3(p(b)) = \mathbf{t} \\
&I_4 : I_4(p(a)) = \bot, \quad I_4(p(b)) = \bot \\
&I_5 : I_5(p(a)) = \bot, \quad I_t(p(b)) = \mathbf{t} \\
&I_6 : I_6(p(a)) = \mathbf{t}, \quad I_6(p(b)) = \bot \\
&I_7 : I_7(p(a)) = \top, \quad I_7(p(b)) = \top \\
&I_8 : I_8(p(a)) = \mathbf{t}, \quad I_8(p(b)) = \top \\
&I_9 : I_4(p(a)) = \top, \quad I_9(p(b)) = \mathbf{t}
\end{aligned}$$

Here, G has a least model (viz. I_4) and a greatest model (viz. I_7). The least model says that from G, it is not *known* that $p(a)$ is true and it is not *known* that $p(b)$ is true. Similarly, I_3 says that if it is *known* that $p(a)$ is false then it is *known* that $p(b)$ is true.

We turn to a fixpoint semantics. Their fixpoint semantics characterizes the models of a program in terms of the fixpoints of a certain monotone operator usually denoted by T_P. x is a pre-fixpoint (post-fixpoint) of a function $f : L \to L$, where L is a lattice under the ordering $\sqsubseteq$ iff $f(x) \sqsubseteq x$ ($x \sqsubseteq f(x)$). x is a fixpoint of f iff $f(x) = x$.

If S is a subset of a complete lattice L, then $\sqcup S$ and $\sqcap S$ denote the least upper bound (lub) of S and the greatest lower bound (glb) of S, respectively.

Definition 6.12 Suppose G is a GHP. Then, T_G is a mapping from the Herbrand interpretations of G to the Herbrand interpretations of G defined by

$T_G(I)(A) = \text{lub}\{\mu \mid A : \mu \Leftarrow B_1 : \mu_1 \& \ldots \& B_k : \mu_k$ is a ground instance of a gh-clause in G and $I \models B_1 : \mu_1 \& \ldots \& B_k : \mu_k\}$.

Example 6.3 Consider the GHP G given below:

$$p(a) : \mathbf{t} \Leftarrow, \quad p(a) : \mathbf{f} \Leftarrow p(a) : \mathbf{t}$$

and the interpretation I that assigns $\mathbf{t}$ to $p(a)$. Then, $T_G(I)(p(a)) = \text{lub}\{\mathbf{t}, \mathbf{f}\} = \top$.

Theorem 6.2 *T_G is monotonic.*

Proof Suppose $I_1 \leq I_2$ where I_1, I_2 are interpretations. Then, if $I_1 \models B : \mu$, then $I_2 \models B : \mu$ for all $B \in B_G$. Thus,

$T_G(I_1)(A)$
$= \text{lub}\{\mu \mid A : \mu \Leftarrow B_1 : \mu_1 \& \ldots \& B_k : \mu_k$ is a ground instance of a gh-clause in G and $I_1 \models B_1 : \mu_1 \& \ldots \& B_k : \mu_k\}$
$\leq \text{lub}\{\mu \mid A : \mu \Leftarrow B_1 : \mu_1 \& \ldots \& B_k : \mu_k$ is a ground instance of a gh-clause in G and $I_2 \models B_1 : \mu_1 \& \ldots \& B_k : \mu_k\}$
$= T_G(I_2)(A)$

Theorem 6.3 states that the pre-fixpoints of T_G are exactly the models of G.

Theorem 6.3 *I is a model of the GHP G iff $T_G(I) \leq I$.*

Proof $T_G(I)(A) \leq I(A)$

iff (directly from the definition of T_G), $\sqcup\{\mu \mid A : \mu \Leftarrow Body$ is a ground instance of a gh-clause in G and $I \models Body\} \leq I(A)$
iff for every ground instance of the form $A : \mu \Leftarrow Body$ of a gh-clause in G, $I \models Body$ implies $I(A) \geq \mu$
iff $I \models A : \mu \Leftarrow Body$ for all ground instances of the form $A : \mu \Leftarrow Body$ of gh-clauses in G
iff $I \models G$.

The monotonicity of T_G assures us that T_G has a fixpoint, and hence pre-fixpoint, and hence a model. In addition, since T_G is monotone and the set of interpretations (over $\mathscr{T}$) is a complete lattice, the least fixpoint and the least pre-fixpoint must coincide. Thus, we have the following result.

Theorem 6.4 *Any GHP G has a least model* $\mathscr{M}_G$. *In addition, this least model is identical to the least fixpoint* $\mathrm{lfp}(T_G)$ *of* T_G.

However, monotonicity by itself does not guarantee that the least fixpoint of T_G is recursively enumerable, but we can show that this is indeed the case.

Definition 6.13 We define the special interpretation Δ to be the interpretation that assigns the value $\perp$ to all members of B_G. Similarly, ∇ assigns the value $\top$ to all members of B_G.

Definition 6.14 The *upward iteration* of T_G is defined as follows:

$$T_G \uparrow 0 = \Delta, \quad T_G \uparrow \alpha = T_G(T_G \uparrow (\alpha - 1)), \quad T_G \uparrow \lambda = \sqcup\{T_G \uparrow \eta \mid \eta < \lambda\}$$

where α is a successor ordinal and λ is a limit ordinal.

The *downarrow iteration* of T_G is defined as follows:

$$T_G \downarrow 0 = \nabla, \quad T_G \downarrow \alpha = T_G(T_G \downarrow (\alpha - 1)), \quad T_G \downarrow \lambda = \sqcap\{T_G \downarrow \eta \mid \eta < \lambda\}$$

where α is a successor ordinal and λ is a limit ordinal.

We observe that $T_G \uparrow 0 \leq T_G \uparrow 1 \leq \cdots \leq T_G \uparrow \omega$ and that $T_G \downarrow 0 \geq T_G \downarrow 1 \geq \cdots \geq T_G \downarrow \omega$.

Theorem 6.5 $T_G \uparrow \omega = \mathscr{M}_G$.

Proof ($\mathscr{M}_G \leq T_G \uparrow \omega$). It is sufficient to show that $T_G \uparrow \omega$ is a model of G. Suppose

$$A : \mu \Leftarrow B_1 : \psi_1 \& \ldots \& B_k : \psi_k$$

is a ground instance $C\theta$ of some gh-clause $C \in G$, and suppose

$$T_G \uparrow \omega \models B_1 : \psi_1 \& \ldots \& B_k : \psi_k.$$

Then, it must be true that for some finite n,

$$T_G \uparrow n \models B_1 : \psi_1 \& \ldots \& B_k : \psi_k.$$

But then, $T_G \uparrow (n+1) = T_G(T_G \uparrow n) \models A : \mu$ (by definition of T_G), hence it follows that $T_G \uparrow \omega \models A : \mu$ by the above remark. Thus, $T_G \uparrow \omega$ is a model of G, and is therefore a pre-fixpoint of T_G.
($T_G \uparrow \omega \leq \mathscr{M}_G$). Since T_G is monotonic, for all α, $T_G \uparrow \alpha \leq \mathscr{M}_G$ where $\mathscr{M}_G$ is the least pre-fixpoint of T_G. Hence, $T_G \uparrow \omega$ is the least pre-fixpoint of T_G.

A subset D of a complete lattice $(L, \sqsubseteq)$ is *directed* iff for every pair of elements a, b in P, there is an element $c \in D$ such that $a \sqsubseteq c$ and $b \sqsubseteq c$.

Definition 6.15 Let $\sqcup S$ be the least upper bound (in L) of the subset S of L. An operator $T : L \to L$ is *weakly continuous* iff

$$T(\sqcup D) \sqsubseteq \{T(d) \mid d \in D\}.$$

T is *continuous* iff

$$T(\sqcup D) = \{T(d) \mid d \in D\}.$$

Lemma 6.2 *$T : L \to L$ is continuous iff T is monotonic and weakly continuous.*

Theorem 6.6 *T_G is continuous.*

Proof Since T_G is monotone, it suffices to prove that T_G is weakly continuous. Recall that the set of interpretations of GHP G is the set of mappings (of type) $B_L \to \mathcal{T}$ and this set is a complete lattice where $I_1 \le I_2$ iff $I_1(A) \le I_2(A)$ for all $A \in B_L$. Let D be a non-empty directed subset of $B_L \to \mathcal{T}$, and let $J = \sqcup D$.

It is necessary to show that $T_G(J) \preceq \sqcup\{T_G(I) \mid I \in D\}$. It suffices to show that for all $a \in B_L$, there is an $I \in D$ such that $T_G(J)(A) = T_G(I)(A)$. This follows at once since $T_G(J)(A) \in \mathcal{T}$ and $\mathcal{T}$ is finite.

The *definite* part of an interpretation I is the set $\{A : \mu \mid I(A) = \mu \text{ and } \mu \neq \perp\}$. The continuity of T_G implies that the definite part of each $T_G \uparrow n$ and the definite part of $T_G \uparrow \omega$ is recursively enumerable by the techniques of recursion theory. Hence, the definite part of the least model of G is recursively enumerable.

Theorem 6.5 raises the following question "Is $T_G \downarrow \omega$ identical to the greatest model?" Unfortunately, the answer to this question is negative. However, the reason for this is that $T_G \downarrow 0$ is the greatest model of G. Indeed, each $T_G \downarrow \alpha$ is a model of G, since it is a pre-fixpoint of T_G. Thus, we answer the following question: "What kind of models would we like GHPs to have, and in addition, what kind of GHPs should be qualified as being acceptable or decent?"

Definition 6.16 An interpretation I of a GHP G is said to be *nice* if the following condition holds:

$$(\forall a \in b_G) I(A) \neq \top$$

We are therefore interested in those models of G that are nice. Since $\top$ denotes both true and false, namely inconsistency, we should like to characterize the *acceptable* GHPs to be exactly those GHPs that possess a nice model. It is appropriate to identify a class $\mathcal{C}$ of GHPs that are guaranteed to possesses nice models. In addition, it is desirable that $\mathcal{C}$ be decidable.

Definition 6.17 A GHP G is *well-behaved* iff the gh-clauses of G satisfying the following condition:

If G_1, G_2 are gh-clauses in G such that their (respective) heads are $A_1 : \mu_1, A_2 : \mu_2$ and if A_1, A_2 are unifiable, then $\mu_1 = \mu_2$.

It remains to prove the claim that every well-behaved GHP has a nice model. We go one step further and show that for well-behave GHPs, $T_G \uparrow \omega$ is nice. We already know that $T_G \uparrow \omega$ is a model of G, so this is sufficient to show our claim.

Lemma 6.3 *G has a nice model iff $T_G \uparrow \omega$ is nice.*

Theorem 6.7 *If G is a well-behaved GHP, then $T_G \uparrow \omega$ is nice.*

Proof Suppose G is well-behaved and $T_G \uparrow \omega$ is not nice. Then, there is some $A \in B_G$ such that $T_G \uparrow \omega(A) = \top$. But then, there exists some finite n such that $T_G \uparrow n(A) = \top$. However, there must exist two gh-clauses C_1, C_2 in G of the form:

$$A_1 : \mathbf{t} \Leftarrow B_1 : \psi_1 \& \ldots \& B_m : \psi_m$$
$$A_2 : \mathbf{f} \Leftarrow D_1 : \rho_1 \& \ldots \& D_k : \rho_m$$

such that A is an instance of A_1, A_2 and $T_G \uparrow (n-1) \models B_1 : \psi_1 \& \ldots \& B_m : \psi_m$ and $T_G \uparrow (n-1) \models D_1 : \rho_1 \& \ldots \& D_k : \rho_k$. As A is a common instance of A_1, A_2, they are unifiable, thus contradicting the assumption that G is well-behaved. Therefore $T_G \uparrow \omega$ must be nice.

Theorem 6.8 *In general, the set*

$$\text{NICE}(G) = \{G \mid G \textit{ is a GHP having at least one nice model}\}$$

is undecidable (and indeed is Π^0_1-complete).

Proof Given recursively enumerable set W, we can find an ordinary definite clause program P such that $T_P \uparrow \omega$ is recursively isomorphic to W; see Andreka and Nemeti [33]. Obtain GHP G from P by annotating all the atoms in P with **t**, and by adding the unit gh-clause $p(a) : \mathbf{f} \Leftarrow$. Then, $T_G \uparrow \omega$ is a nice model of G iff $p(a) \notin T_P \uparrow \omega$. It follows that the problem of whether $T_G \uparrow \omega$ is a nice model of G by standard techniques of recursion theory. By Lemma 6.3, the problem of whether G has a nice model is Π^0_1-complete.

We can define recursive subsets of NICE(G). Theorem 6.8, together with the fact that $T_G \uparrow \omega$ is recursively enumerable, entails that the set NICE(G) is Π^0_1-complete.

Theorem 6.9 *If G is a well-behaved GHP, then $T_G \downarrow 1$ is nice; hence, $T_G \downarrow \alpha$ is nice for all $\alpha > 0$.*

Proof Suppose G is well-behaved and $T_G \downarrow 1$ is not nice. Then, there is some $A \in B_G$ such that $T_G \downarrow 1(A) = \top$. Then, there must be at least two gh-clauses C_1, C_2 in G having ground instances of the following form:

$$A : \mathbf{t} \Leftarrow D_1 : \rho_1 \& \ldots \& D_k : \rho_k$$
$$A : \mathbf{f} \Leftarrow E_1 : \psi_1 \& \ldots \& E_r : \psi_k$$

such that $T_G \downarrow 0$ satisfies the bodies of both the above instances of C_1 and C_2. But, the heads of C_1, C_2 are unifiable and the annotations of the heads are distinct, thus contradicting the well-behavedness of G. Therefore, $T_G \downarrow 1$ is nice.

Example 6.4 Consider the well-behaved GHP G given below:

$$p(a) : \mathbf{t} \Leftarrow q(X) : \mathbf{t}, \quad q(s(X)) : \mathbf{t} \Leftarrow q(X) : \mathbf{t}$$

In this case, $T_G \uparrow \omega$, while $T_G \downarrow \omega(p(a)) = \mathbf{t}$. For every ground atom $A \in B_G$, $A \neq p(a)$, $T_G \downarrow \omega(A) = \bot$. However, $T_G \downarrow \omega(p(a)) = \mathbf{t}$. Thus, $T_G \downarrow \omega$ is not a fixpoint of T_G as $T_G \downarrow (\omega + 1)(p(a)) = \bot$.

Example 6.4 illustrates the fact that while $T \uparrow \omega$ is a fixpoint of T_G, $T_G \downarrow$ may not be a fixpoint of T_G.

Definition 6.18 A model I of a GHP G is said to be *supported* if for every ground atom $A \in B_G$ such that $I(A) \neq \bot$, there is a gh-clause in G having a ground instance of the form:

$$A : \mu \Leftarrow B_1 : \psi_1 \& \ldots \& B_m : \psi_m$$

such that $I \models B_1 : \psi_1 \& \ldots \& B_m : \psi_m$ and $I(A) = \mu$.

Note that all literals occurring in a GHP are well annotated. Hence, $\mu \neq \top$. Thus, every supported model must be nice.

We now comment on the relation of the fixpoints of T_G and the supported models of G.

Lemma 6.4 *If I is a supported (and hence, nice) model of the GHP G, then I is a fixpoint of T_G.*

Proof Suppose I is a supported model of G and $I(A) = \mu$ and $\mu \notin \{\bot, \top\}$. Let $\Gamma = \{A : \rho \mid$ there is a ground instance C of a gh-clause in G such that C is $A : \rho \Leftarrow F$ and $I \models F\}$. $A\mu \in \Gamma$ since I is a supported model of G. Since $I \models G$, $\mu = I(A) \leq \sqcup\{\rho \models A : \rho \in \Gamma\} = T_G(I)(A)$. But $T_G(I)(A) \geq \mu$ since $\mu \in \{\rho \mid A : \rho \subset \Gamma\}$. Therefore, $\mu = I(A) = T_G(I)(A)$.

Lemma 6.5 *If I is a nice fixpoint of T_G, then I is a supported model of G.*

Proof Assume $I(A) \neq \bot$. As I is nice, $I(A)$ must be in $\{\mathbf{t}, \mathbf{f}\}$.
Case 1: $[I(A) = \mathbf{t}]$ Since I is a fixpoint of T_G, $T_G(A)(A) = \mathbf{t}$. That is, by definition, $\sqcup\{\mu \mid A : \mu \Leftarrow B_1 : \psi_1 \& \ldots \& B_k : \psi_k$ is a ground instance of a gh-clause in G and $I \models B_1 : \psi_1 \& \ldots \& B_k : \psi_k\} = \mathbf{t}$. But for the lub above to be $\mathbf{t}$, there must be a gh-clause in G having a ground instance of the $A : \mathbf{t} \Leftarrow B_1 : \psi_1 \& \ldots \& B_k : \psi_k$ and such that $I \models B_1 : \psi_1 \& \ldots \& B_k : \psi_k$, i.e., I is supported.
Case 2: $[I(A) = \mathbf{f}]$ Similar to Case 1.

Theorem 6.10 *I is a supported model of the GHP G iff I is a nice fixpoint of G.*

Jaffar et al. [93] introduced a decency thesis in conventional logic programming, which claims that all decent logic programs are *canonical*, i.e., $T_P \downarrow \omega = \text{gfp}(T_P)$ where T_P is defined as in Lloyd [111]. The term decent is intended to refer to programs that arise "naturally" in good practice.

Definition 6.19 A GHP G is *canonical* iff $T_G \downarrow \omega$ is a fixpoint of T_G.

It follows immediately from Lemma 6.5 that if G is well-behaved canonical program, then $T_G \downarrow \omega$ is a supported model of G.

Theorem 6.11 *$T_G \downarrow \omega$ is the greatest (nice) supported model of G if G is a canonical well-behaved GHP.*

Proof Since G is canonical, $T_G \downarrow \omega$ is a fixpoint of T_G, which must be the greatest fixpoint of T_G as T_G is monotonic. Since G is well-behaved, $T_G \downarrow \omega$ is nice. Thus, $T_G \downarrow \omega$ is the greatest supported model of G.

Next, we describe an operational semantics of GHPs. First, an and-or tree style procedure, which is similar to the approach of van Emden [153], is given. Second, we give an exposition of the resolution style procedure.

Definition 6.20 Suppose G is a GHP, A is a (not necessarily ground) atom in the language of G, and μ an annotation. We define an and/or tree $\mathbf{T}(G, A : \mu)$ as follows:

1. The root of $\mathbf{T}(G, A : \mu)$ is an or-node labelled $a : \mu$.
2. If N is an or-node, then it is labelled by a single annotated literal.
3. Each and-node is labelled by a gh-clause from G and substitution.
4. Descendants of an or-node are and-nodes and descendents of and-nodes are or-nodes.
5. If N is an or-node labelled by $B : \alpha$ ($\alpha \neq \perp$), and if $C\theta$ is an instance of a gh-clause C in G of the following form:

   $$B : \beta \Leftarrow D_1 : \psi_1 \& \ldots \& B_k : \psi_k$$

   where $\beta \geq \alpha$, then there is a descendent of N labelled by C and θ. An or-node with no descendants is called the *uninformative* node.
6. If N is an and-node labelled by the gh-clause C and the substitution θ, then for every annotated literal $B : \gamma$ in the body of C, there is a descendant or-node labelled $B\theta : \gamma$. An and-node with no descendents is called a *success* node.

Associated with every node N in the and/or tree $\mathbf{T}(G, A, \mu)$ is a truth-value $v(N)$ called the value of that node. v is defined as follows:

If N is a success node labelled $B : \psi$, then $v(N) = \psi$.

If N is an uninformative node, then $v(N) = \perp$.

If N is an or-node that is not uninformative and its descendants are $N_1, \ldots, N_k$ then $v(N) = \text{lub}\{v(N_1), \ldots, v(N_k)\}$.

If N is a non-terminal and-node labelled with the gh-clause $A : \rho \Leftarrow B_1 : \psi_1 \& \ldots \& B_k : \psi_k$, and if the value $v(N_i)$ of each of its descendant nodes N_i labelled B_i is such that $v(N_i) \geq \psi_i$ for all $1 \leq i \leq k$, then $v(N) = \rho$, else $v(N) = \perp$.

Definition 6.21 A GHP G is said to be *covered* if every variable symbol that occurs in the body of a gh-clause $C \in G$ occurs in the head of C.

Theorem 6.12 *If G is a covered GHP, $A \in B_G$ and if $\mathscr{T}(G, A : \mu)$ is finite with root R, then $v(R) \leq T_G \uparrow \omega(A)$.*

Proof Similar to the proof of Theorem 3.1 of van Emden [153]. Note that his proof applies only to covered rule sets.

Theorem 6.13 *If G is a GHP, $A \in B_G$, and if $\mathscr{T}(G, A : \mu)$ is finite with root R, then $v(R) \geq T_G \uparrow \omega(A)$.*

Proof Let R be the root of $\mathscr{T}(G, A : \mu)$. It suffices to prove by induction that

$$(\forall n \in N) v(R) \geq T_G \uparrow n(A)$$

where N denotes the set of non-negative integers.

Base case ($n = 0$): Trivially true as $T_G \uparrow 0(A) = \perp$.

Inductive case: Let $\alpha = T_G \uparrow (n + 1)(A)$.

Case 1 ($\alpha = \top$): Then, there are two ground instances $C_1\theta_1, C_2\theta$ of gh-clauses $C_1, C_2 \in G$ of the form:

$$A : \mathbf{t} \Leftarrow B_1 : \psi_1 \& \ldots \& B_r : \psi_r,$$
$$A : \mathbf{f} \Leftarrow D_1 : \rho_1 \& \ldots \& D_s : \rho_s$$

and $T_G \uparrow n$ satisfies the bodies of $C_1\theta_1$ and $C_2\theta_2$, respectively. each B_i and D_j is ground, and therefore, by the induction hypothesis, $v(B'_i) \geq \psi_i$ and $v(D'_j) \geq \rho_j$ where B'_i and D'_j are the atoms in the roots of the and/or trees $\mathscr{T}(G, B_i : \psi_i)$ and $\mathscr{T}(G, D_j, \rho_j)$, respectively. Therefore, the descendant nodes N_{C_1}, N_{C_2} of R that are labelled with C_1, C_2, respectively, are such that $v(N_{C_1}) \geq \mathbf{t}$, $v(N_{C_2}) \geq \mathbf{f}$, respectively, whence $v(R) \geq \text{lub}\{\mathbf{t}, \mathbf{f}\} = \top$. This completes this case.

Case 2 ($\alpha = \perp$): Trivial.

Case 3 ($\alpha = \mathbf{t}$ or $\mathbf{f}$): In this case, the analysis is almost identical to that of the case when $\alpha = \top$ except that we consider only one gh-clause C_1 instead two.

Corollary 6.1 (Soundness) *If R is a covered GHP, $A \in B_G$ and $\mathscr{T}(G, A : \mu)$ is finite, then $v(R) = T_G \uparrow \omega(A)$ where R is the root of $\mathscr{T}(G, A : \mu)$.*

As Blair and Subrahmanian [53] pointed out, it is not possible to give an SLD-resolution like procedure for GHPs. However, we can define an SLD-resolution like procedure, called the *SLDnh-resolution*, for well-behaved GHPs.

Recall that a literal is well annotated if the annotation is either **t** or **f**. A *query* is the existential closure of a conjunction of w-annotated literals. We will often formulate queries as a conjunction of w-annotated literals and assume all variables in it to be implicitly existentially quantified.

Definition 6.22 (*nh-resolvent*) An *nh-resovent* with respect to $A_i : \rho_i$ of the query Q given by $A_1 : \rho_1 \& \ldots \& A_k : \rho_k$ and the gh-clause C of the form $D : \beta \Leftarrow B_1 : \psi_1 \& \ldots \& B_r : \psi_r$ is the query

$$(A_1 : \rho_1 \& \ldots \& A_{i-1}\rho_{i-1} \& B_1 : \psi_1 \& \ldots \& B_r : \psi_r \& A_{i+1} : \rho_{i+1} \& \ldots \& A_k : \rho_k)\theta$$

where $\beta \geq \rho_i$ and θ is the most general unifier of D and A_i. We here assume that C and Q contain no variables in common. $A_i : \rho_i$ and $D : \rho$ are called the literals nh-resolved upon ("nh" stands for "non-Horn"). Note that an nh-resolvent is always a query.

Definition 6.23 (*SLDnh-deduction*) An *SLDnh-deduction* from the *initial* query Q_0 and the GHP G is a sequence

$$\langle Q_0, C_1, \theta_1\rangle, \ldots, \langle Q_i, C_{i+1}, \theta_{i+1}\rangle, \ldots$$

where Q_{r+1} is the nh-resolvent of Q_r and C_{r+1} and C_{r+1} is a renamed version of gh-clause in G such that C_{r+1} has no variable symbols in common with any of $Q_0, \ldots, Q_r, C_1, \ldots, C_r$, and θ_{r+1} is the most general unifier of the literals nh-resolved upon.

Definition 6.24 (*SLDnh-refutation*) An *SLDnh-refutation* of the initial query q_0 is a finite SLDnh-deduction

$$\langle Q_0, C_1, \theta_1\rangle, \ldots, \langle Q_i, C_{i+1}, \theta_{i+1}\rangle$$

such that the result of resolving Q_n and Q_{n+1} is the Q_{n+1} is the empty gh-clause.

Theorem 6.14 (Soundness) *If there exists an SLDnh-refutation of the initial query* Q_0

$$A_1 : \rho_1 \& \ldots \& A_m : \rho_m$$

from the GHP G, then

$$T_G \uparrow \omega \models \exists(Q_0)$$

Proof Suppose there exists an SLDnh-refutation of the initial query Q_0 from G of the following form:

$$\langle Q_0, C_1, \theta_1\rangle, \ldots, \langle Q_n, C_{n+1}, \theta_{n+1}\rangle$$

We will show by induction on n that $T_G \uparrow \omega \models \exists(Q_0)$.

Base case ($n = 0$): then, the nh-resolvent of Q_0 and C_1 is the empty query, i.e., Q_0 is a unit conjunction, i.e., $A_1 : \rho_1$, and C_1 is a unit gh-clause, i.e., C_1 is the form $A' : \beta \Leftarrow$ where $\beta \geq \rho_1$ and $A_1\theta_1 = A'\theta_1$. As $T_G \uparrow \omega$ is a model of the GHP G, it must be a model of C_1, and so it must be true that for every ground instance A_2 of A', $T_G \uparrow \omega(A_2) \geq \beta$. In particular, for every A_3 that is a ground instance of $A'\theta_1$, $T_G \uparrow \omega(A_3) \geq \beta \geq \rho_1$, whence $T_G \uparrow \omega \models \exists Q_0$.

Induction case: Suppose

$$\langle Q_0, C_1, \theta_1\rangle, \ldots, \langle Q_n, C_{n+1}, \theta_{n+1}\rangle, \langle Q_{n+1}, C_{n+2}, \theta_{n+2}\rangle$$

is an SLDnh-refutation of Q_0. Then

$$\langle Q_0, C_1, \theta_1\rangle, \ldots, \langle Q_n, C_{n+1}, \theta_{n+1}\rangle$$

is an SLDnh refutation of Q_1. Therefore, by induction hypothesis, $T_G \uparrow \omega \models \exists(Q_1)$. But Q_1 is the nh-resolvent of Q_0 and C_1. So Q_1 is of the form

$$(A_1 : \rho_1 \& \ldots \& A_{i-1} : \rho_{i-1} \& E_1 : \psi_1 \& \ldots \& E_r : \psi_r \& A_{i+1} : \rho_{i+1} \& \ldots \& A_m : \rho_m)\theta_1$$

where C_1 is the gh-clause

$$H : \delta \Leftarrow E_1 : \psi_1 \& \ldots \& E_r : \psi_r$$

such that $H\theta_1 = A_i\theta_1$ and $\delta \geq \rho_i$ (b definition of nh-resolution). As $T_G \uparrow \omega \models (\exists)Q_1$, it follows that

$$T_G \uparrow \omega \models (E_1 : \psi_1 \& \ldots \& E_r : \psi_r)\theta_1$$

(as this is a sub-conjunct of Q_1). as $T_G \uparrow \omega$ is a model of G (Theorem 6.5), it must satisfy every gh-clause of G (and every renamed version of any gh-clause in G). in particular, $T_G \uparrow \omega$ must satisfy $C_1\theta_1$. As $T_G \uparrow \omega$ satisfies the body of $C_1\theta_1$, it must satisfy $H\theta_1 : \delta$. But $H\theta_1 = A_i\theta_1$; so $T_G \uparrow \models A_i\theta_1 : \rho_i$. Thus,

$$T_G \uparrow \omega \models (\exists)(A_1 : \rho_1 \& \ldots \& A_m : \rho_m)\theta_1$$

Notice that the soundness theorem does not restrict the GHP to covered GHPs. However, this restriction is needed for our completeness result.

Theorem 6.15 (Completeness) *Suppose G is covered, well-behaved GHP. Then if* $Q_0 = A_1 : \rho_1 \& \ldots \& A_m : \rho_m$ *is a ground query that is satisfied by* $T_G \uparrow \omega$, *then there is an SLDnh-refutation of* Q_0 *from G.*

Proof Suppose G is covered and well behaved. Let $Q_0 = (A_1 : \rho_1 \& \ldots \& A_m : \rho_m)$ be a ground query that is satisfied by $T_G \uparrow \omega$. Then, there must be some $n > 0$ (as $\rho_i \neq \bot$ for all i) such that $T_G \uparrow n \models Q_0$. We will prove by induction n that an SLDnh-refutation of Q_0 from G exists.

Base case: Suppose $T_G \uparrow 1 \models Q_0$. Then for each $1 \leq i \leq m$, there is a gh-clause C_i in G having a ground instance $C_i\theta_i$ of the form $A : \beta_i \Leftarrow$ where $\beta_i \geq \rho_i$. As Q_0 is ground, none of the substitution θI affect it, and so

$$\begin{array}{l}\langle A_1 : \rho_1 \& \ldots \& A_m : \rho_m, C_1, \theta_1 \rangle \\ \langle A_2 : \rho_2 \& \ldots \& A_m : \rho_m, C_2, \theta_2 \rangle \\ \vdots \\ \langle A_m : \rho_m \& \ldots \& A_m : \rho_m, C_m, \theta_m \rangle\end{array}$$

is an SLDnh-refutation of Q_0.

Inductive case: Suppose $T_G \uparrow (r+1) \models A_1 : \rho_1 \& \ldots \& A_m : \rho_m$. Then, for each A_i, there is a gh-clause C_i in G having a ground instance $C_i\theta_i$ of the form

$$A_i : \beta_i \Leftarrow B_1^i \& \ldots \& B_{k_i}^i$$

where $\beta_i \geq \rho_i$, and $T_G \uparrow r \models B_1^i \& \ldots \& B_{k_i}^i$. Now, as each A_i is ground, and as G is covered, $B_1^i \& \ldots \& B_{k_i}^i$ is ground and so, by induction hypothesis, there exists an

SLDnh-refutation $\mathbf{R}_i$ of $B_1^i \& \ldots \& B_{k_i}^i$. Therefore,

$$\langle A_i : \rho_i, C_i, \theta_i \rangle, \mathbf{R}_i$$

is a SLDnh-refutation of A_i. Denote the SLDnh-refutation tree of each $A_i : \rho_i$ by Γ_i. This is exactly an SLDnh-refutation tree for Q_0.

Later, Kifer and Lozinskii [98, 99] extended the theory of paraconsistent logic programming to a full-fledged logic, and showed that a sound and complete proof procedure exists. Da Costa et al. [64] studied theorem-proving of annotated logic, presenting theory and implementation.

6.2 Generalized Annotated Logic Programming

Since Blair and Subrahmanian's theory is based on the special lattice *FOUR*, it is expected to extend the theory based on other types of lattices. Kifer and Subrahmanian [100] established the issue by proposing *generalized annotated logic programming* (*GAP*).

Here, we give an exposition of generalized annotated logic programming following [101]. We assume an upper semilattice $\mathcal{T}$ of truth-values with the semilattice ordering $\leq$ and we denote the least upper bound operator by $\sqcup$. The semilattice needs not be complete. Some of the results in [100] will depend on the existence of a greatest element denoted $\top$ in $\mathcal{T}$.

The greatest lower bound operator is denoted by $\sqcap$. Elements of $\mathcal{T}$ can be thought of as confidence factors or degrees of belief. In addition, we will sometimes assume that $\mathcal{T}$ has a unique least element denoted $\perp$.

For each $i \geq 1$, we postulate that there is a family $\mathcal{F}_i$ of total continuous (hence monotonic) functions, each of type $\mathcal{T}^i \to \mathcal{T}$, called *annotated functions*. We denote $\mathcal{F} = \cup_{i \geq 1} \mathcal{F}_i$ and assume that all functions f in $\mathcal{F}$ are computable. We also assume that each $\mathcal{F}_j$ contains a j-ary function $\sqcup_j$, derived from the semilattice operator $\sqcup$, which, given inputs $\mu_1, \ldots, \mu_j$ returns the least upper bound of $\{\mu_1, \ldots, \mu_j\}$.

The language of GAP is first-order language for logic programs with annotation functions. We also postulate two disjoint sets of variable symbols, i.e., *object variables* and *annotation variables*.

Definition 6.25 (*Formulas*) An *annotation* is either an element of $\mathcal{T}$, an *annotation variable*, or a *complex annotation term*. Annotation terms are defined recursively as follows: member of $\mathcal{T}$ and annotation variables are annotation terms. In addition, if $f \in \mathcal{F}_n$ and $x_1, \ldots, x_n$ are annotation terms, then $f(x_1, \ldots, x_n)$ is a complex annotation term.

If A is a usual atomic formula of first-order logic and α is an annotation, then $A : \alpha$ is an *annotated atom*. An annotated atom containing no occurrence of object variables is *ground*. If $\alpha \in \mathcal{T}$ then $A : \alpha$ is *constant-annotated* (or *c-annotated*). When α is an annotation variable, then $A : \alpha$ is said to be *variable-annotated* (or *v-annotated*). If α is a complex annotation term, then $A : \alpha$ is *term-annotated* (*t-annotated*).

Definition 6.26 (*Annotated clause*) If $A : \rho$ is an annotated atom and $B_1 : \mu_1, \ldots, B_k : \mu_k$ are c- or v-annotated atoms, then

$$A : \rho \leftarrow B_1 : \mu_1, \ldots, B_k : \mu_k$$

is an *annotated clause*. $A : \rho$ is called the *head* of this clause, while $B_1 : \mu_1, \ldots, B_k : \mu_k$ is called the *body*. All variables (object or annotation) are implicitly universally quantified.

Any set of annotated clauses is called a *generalized annotated program* (*GAP*). By the definition, members of $\mathcal{F}$ may occur in the annotation of head of a clause, but elements of $\mathcal{F}$ may not occur in the body of a clause.

Definition 6.27 Suppose C is an annotated clause. A *c-annotated instance* of C is any annotated clause obtained by replacing all annotation variables occurring in C by members of $\mathcal{T}$. Different occurrences of the same annotation variable must be replaced by the same members of $\mathcal{T}$.

Definition 6.28 Suppose C is an annotated clause. A *strictly ground instance* of C is any ground instance of C which contains only c-annotations. Notice that since all functions in $\mathcal{F}$ are evaluable, annotation terms of the form $f(a_1, \ldots, a_n)$, where $a_1, \ldots, a_n \in \mathcal{T}$ and $f \in \mathcal{F}_n$ are also considered to be ground and are identified with the result of the computation of f on the a_i's.

We denote by *SGI*(C) the set of all strictly ground instances of a clause C and *SGI*(P) the set of all strictly ground instances of clauses in a *GAP* P.

Kifer and Subrahmanian [100] showed two semantics for GAPs, i.e., restricted and general semantics. A general semantics was studied by Kifer and Lozinskii [98, 99], and both restricted semantics and general semantics were extensively investigated by Kifer and Subrahmanian [100].

Definition 6.29 An *ideal* of an upper semilattice is any subset S such that:

1. S is *downward closed*, i.e., $s \in S$ and $t \leq s$ imply $t \in S$
2. S is closed with respect to *finite* least upper bounds, i.e., $s, t \in S$ implies $s \sqcup t \in S$.

an ideal S is *principal* if from some $p \in \mathcal{T}$, $S = \{s \mid s \leq p\}$. S is called the principal ideal *generated* by p and is denoted by $\| p \|$. The set of all ideals of $\mathcal{T}$ is denoted by $\mathcal{I}(\mathcal{T})$, and the set of principal ideals of $\mathcal{T}$ by $\mathcal{PI}(\mathcal{T})$.

Example 6.5 Ideals are not necessarily closed under *infinite* least upper bounds. For example, consider the complete lattice $[0, 1]$ (the unit interval of reals) ordered by $\leq$. Then, the right-open interval

$$[0, 1) = \{x \mid 0 \leq x < 1\}$$

is an ideal which is not closed under infinite least upper bound operator.

Observe that $\mathscr{I}(\mathscr{T})$ forms a complete lattice with the intersection operation serving as the greatest lower bound and the union operator serving as the least upper bound; the order on $\mathscr{I}(\mathscr{T})$ is determined by the usual set inclusion $\subseteq$. Furthermore, there is a homomorphic embedding of upper semilattice $\mathscr{T} \mapsto \mathscr{I}(\mathscr{T})$ (that preserves finite least upper bounds) that maps elements of $\mathscr{T}$ into the corresponding principal ideals of $\mathscr{I}(\mathscr{T})$.

Definition 6.30 (*General and Restricted Herbrand interpretations*) Let $\mathscr{L}$ be a language of annotated logic. The *Herbrand base* of $\mathscr{L}$, $B_{\mathscr{L}}$, is the set of all ground atomic formulas of $\mathscr{L}$ (without annotations).

A *general Herbrand interpretation* (or interpretation) I is a mapping from the Herbrand base of $\mathscr{L}$ to $\mathscr{I}(\mathscr{L})$. Since I is a function into a partially ordered set $\mathscr{I}(\mathscr{T})$, we can define a partial order on interpretations in the usual way: $I \leq J$ iff for every $p \in B_{\mathscr{L}}$, $I(p) \subseteq J(p)$.

A *restricted Herbrand interpretation* (*r-interpretation*) of $\mathscr{L}$ is any map from $B_{\mathscr{L}}$ to $\mathscr{T}$. Equivalently, a r-interpretation of $\mathscr{L}$ is a map from $B_{\mathscr{L}}$ to $\mathscr{P}\mathscr{I}(\mathscr{T})$, the set of *principal* ideals of $\mathscr{T}$.

There is a distinction between these interpretations. Restricted Herbrand interpretations assign a single truth-value, i.e., essentially a *principal* ideal to ground atoms, whereas general Herbrand interpretations assign *arbitrary* ideals to atoms. Therefore, every r-interpretation is also a general interpretation, but not vice versa.

We also assume that there is a unary operator $\neg : \mathscr{T} \rightarrow \mathscr{T}$, which corresponds to *negation*. Note here that we do not impose any restrictions on $\neg$.

Definition 6.31 (*Satisfaction*) Suppose I is a general interpretation. $\mu \in \mathscr{T}$ is a c-annotation in $\mathscr{T}$ and A is a ground atom., Then,

1. If A is a ground atom, $I \models A : \mu$ iff $\mu \in I(A)$, where $\mu \in \mathscr{T}$.
2. $I \models \neg A : \mu$ iff $\neg(\mu) \in I(A)$.
3. $I \models F_1 \& F_2$ iff $I \models F_1$ and $I \models F_2$.
4. $I \models F_1 \vee F_2$ iff $I \models F_1$ or $I \models F_2$.
5. $I \models F_1 \leftarrow F_2$ iff $I \models F_1$ or $I \not\models F_2$.
6. $I \models F_1 \leftrightarrow F_2$ iff $I \models (F_1 \leftarrow F_2)$ and $I \models (F_2 \leftarrow F_1)$.
7. $I \models (\forall x)F$ iff $I \models F(x/t)$ for all ground terms t. Here, x is an object or annotation variable and t must be of the same sort as x (i.e., either a usual round first-order term, or an element of $\mathscr{T}$. $F(x/t)$ denotes the replacement of all free occurrences of x in F by t.
8. $I \models (\exists x)F$ iff $I \models F(x/t)$ for some ground terms t. Here, x is an object or annotation variable.
9. F is not a closed formula, then $I \models F$ iff $I \models (\forall)F$, where $(\forall)F$ denotes the universal closure of F.

Definition 6.32 (*r-satisfaction*) Suppose I is an r-interpretation. $\mu \in \mathscr{T}$ is a c-annotation in $\mathscr{T}$ and A is a ground atom. Then,

1. If A is a ground atom. $I \models^r A : \mu$ iff $I(A) \geq \mu$.
2. $I \models^r \neg A : \mu$ iff $\neg(\mu) \leq I(A)$

The remaining cases (3)–(10) are defined as for general satisfaction.

An interpretation I (or r-interpretation J) is said to be a *model* (*r-model*) of a formula F iff $I \models F$ ($I \models^r F$). I is a model of a set of formulas P iff it is a model of each of the formulas in P. If P is a set of formulas and φ is a formula, we write $P \models \varphi$ ($P \models^r \varphi$) iff whenever $I \models P$ ($I \models^r P$) then $I \models \varphi$ ($I \models^r \varphi$).

Kifer and Lozinskii [98, 99] considered two different notions of negation. One is *ontological negation*, denoted $\sim$, which is close to the standard negation in classical first-order logic. The other is, as Definition 6.31(2), *epistemic negation*, denoted $\neg$, which is close to the negation in multi-valued logic. We can define a satisfaction relation of ontological negation as follows:

$I \models \sim A$ iff $I \not\models A : \mu$

By ontological negation, implication $A \leftarrow B$ can be defined as $A \vee \sim B$. However, $\leftarrow$ cannot be expressed by $\&, \vee$ and $\neg$. Epistemic negation and ontological negation in *GAP* roughly correspond to negation $\neg$ and strong negation $\neg_*$ in $P\tau$ and $Q\tau$ as discussed in Chaps. 2 and 3.

Kifer and Subrahmanian assumed that in every clause, variables occurring in the annotation of the clause head also appears as annotations of the body literals.

Definition 6.33 Suppose I is an interpretation and $A \in B_{\mathscr{L}}$. Then, $T_P(I)(A) =$ the *least* ideal of $\mathscr{T}$ containing the set $\{f(\mu_1, \ldots, \mu_n) \mid A : f(\mu_1, \ldots, \mu_n) \leftarrow B_1 : \mu_1 \& \ldots \& B_n : \mu_n$ *is in SGI*(P), and $I \models (B_1 : \mu_1 \& \ldots \& B_n : \mu_n)\}$. It is easy to see that intersection of an arbitrary number of ideals is an ideal, and therefore for every subset $S \subseteq \mathscr{T}$ there is a unique least ideal containing S. Note also that $T_P(I)(A)$ is a subset, not an element, of $\mathscr{T}$.

Definition 6.34 Suppose I is an interpretation and $A \in B_{\mathscr{L}}$. Assume also that $\mathscr{T}$ is a complete semilattice. Then, $R_P(I)(A) = \sqcup\{f(\mu_1, \ldots, \mu_n) \mid A : f(\mu_1, \ldots, \mu_n) \leftarrow B_1 : \mu_1 \& \ldots \& B_n : \mu_n$ is in *SGI*(P), $I \models (B_1 : \mu_1 \& \ldots \& B_n : \mu_n)\}$.

Notice that if I is an r-interpretation (hence also an interpretation), then $R_P(I)(A) = \sqcup T_P(I)(A)$ for each atom A, where $\sqcup$ is the least upper bound operator with the postulate that $\sqcup\{\ \} = \bot$.

One intuition behind the ideal-theoretic definition is the following: Consider an interpretation I and a ground atom A. If there is a clause in *SGI*(P) with head $A : \mu$ and whose body is satisfied by I, then we may use I to "conclude" that there is a derivation of $A : \mu$ by using *modus ponens*. The difference between R_P and T_P is that R_P would allow *infinite* lubs to be present.

Theorem 6.16 *Suppose P is a GAP. I is an interpretation and J is an r-interpretation. Then,*

1. *I is a model of P iff $T_P(I) \leq I$;*
2. *J is an r-model of P iff $R_P(J) \leq J$;*
3. *T_P is monotonic;*
4. *R_P is monotonic.*

Proof They can be proved by a simple modification of the semantics for logic programs (cf. Lloyd [111]) with the monotonic property of annotated functions in $\mathscr{F}$.

In what follows, we will often use a special "least" interpretation Δ which assigns the empty ideal $\{\ \}$ to every atom. In case of restricted interpretations, the least r-interpretation may not exist, unless we require $\mathscr{T}$ to have the least element $\perp$. In the latter case, the least r-interpretation, denoted Δ_r, assigns $\perp$ to every atom in $B_{\mathscr{L}}$.

Now, we define the iterations of T_P as follows: $T_P \uparrow 0 = \Delta$. If α is a successor ordinal, then $T_P \uparrow \alpha = T_P(T_P \uparrow (\alpha - 1))$; if α is a limit ordinal, then $T_P \uparrow \alpha = \sqcup_{\beta<\alpha} T_P \uparrow \beta$. The iterations of R_P are defined similarly with the exception that $R_P \uparrow 0 = \Delta_r$.

We will see that as in classical logic programming, T_P is continuous, and the equitation $T_P \uparrow \omega = \text{lfp}(T_P)$ holds, but this is not always the case with R_P.

Theorem 6.17 *Let P be a GAP, Then,*

1. *T_P is continuous;*
2. *$T_P \uparrow \omega = \text{lfp}(T_P) =$ the least model of P;*
3. *For all annotated ground atoms $A : \mu$, $P \models A : \mu$ iff $\mu \in T_P \uparrow \omega(A)$.*

Proof The only non-obvious thing is (1). Other claims follow from continuity in a standard way.

To show continuity, let $I_1, I_2, \ldots$ be a directed sequence of interpretations of P (i.e., every finite subsequence $I_{i_1}, \ldots, I_{i_k}$ has an upper bound $I_l : I_l \geq I_{i_j}, j = 1, \ldots, k$). We have to show that $T_P(\sqcup I_i) = \sqcup(T_P(I_i))$. It is easily seen from the definition that for any set of interpretations, $\{J_k\}$, their least upper bound $\sqcup J_k$, is such an interpretation J that for every ground atom A. $J(A)$ is the least ideal containing the set $\cup J_i(A)$.

Since T_P is monotonic, $T_P(I_k) \leq T_P(\sqcup I_i)$ for all k. Since, for every A, $T_P(\sqcup I_i)(A)$ is an ideal, we conclude that $T_P(\sqcup I_i) \geq \sqcup(T_P(I_i))$.

In the other direction, let A be a ground atom such that $\mu \in T_P(\sqcup I_i)(A)$. Then, there must be a strict ground instance of a rule in P of the form: $A : f(\mu_1, \ldots, \mu_n) \leftarrow B_1 : \mu_1 \& \ldots \& B_n : \mu_n$, where $\mu = f(\mu_1, \ldots, \mu_n)$ and the literals $B_j : \mu_j$ are satisfied by $\sqcup I_i$. This means that for every $j = 1, \ldots, n$, there are $\nu_{j1}, \ldots, \nu_{jk_j}$ such that

1. each of the $B_j : \nu_{jl}$ is in some I_i,
2. $\mu_j = \sqcup\{\nu_{j1}, \ldots, \nu_{jk_j}\}$

Since the set I_i of interpretations is directed, there is some I_{i0} that satisfies all the $B_j : \nu_{jl}$. Because of (1) above, $I_{i0} \models B_1 : \nu_{1m_1} \& \ldots \& B_n : \nu_{nm_n}$, for any $m_1, \ldots, m_n$ such that $1 \le m1 \le k_1, \ldots, 1 \le m_n \le k_n$. Hence,

$$f(\nu_{1m_1}, \ldots, \nu_{nm_n}) \in T_P(I_{i0})(A). \qquad (*)$$

Therefore, by continuity of f (all annotation functions are continuous, by definition),

$$\begin{aligned} A : f(\mu_1, \ldots, \mu_n) &= A : f(\sqcup\{\nu_{11}, \ldots, \nu_{1k_1}\}, \ldots, \sqcup\{\nu_{n1}, \ldots, \nu_{nk_n}\}) \\ &= A : \sqcup f(\nu_{1m_1}, \ldots, \nu_{nm_n}) \end{aligned}$$

Thus, because of Eq. (*) and since $T_P(I_{i0})$, being an ideal, is closed under finite least upper bounds, we can conclude that $f(\mu_1, \ldots, \mu_n) \in \sqcup(T_P(I_i))(A)$.

Corollary 6.2 *If A is a ground atom such that $\mu \in T_P \uparrow \omega(A)$, then there is an integer n such that $\mu \in T_P \uparrow n(A)$.*

Proof Since $T_P \uparrow n \subseteq T \uparrow (n+1)$ for all $n \ge 0$, it follows that $(T_P \uparrow \omega)(A) = \cup(T_P \uparrow n)(A)$ (i.e., a plain union of sets instead of the least upper bound $\sqcup$). Therefore, μ must belong to one of the $T_P \uparrow n(A)$'s.

Since T_P is continuous, it attains a fixpoint by the ωth step of upward iterations. But it is not the case for R_P. Blair and Subrahmanian [53] showed that $\text{lfp}(R_P) = R_P \uparrow \omega$ whenever P is c-annotated and $\mathcal{T}$ is lattice.

Kifer and Subrahmanian [100] identified a large class of programs for which R_P is not necessarily continuous, but still $R_P \uparrow \omega = \text{lfp}(R_P)$ holds.

Now, we turn to an SLD-style proof theory for GAPs based on the general semantics.

Definition 6.35 Suppose P is a GAP and $C_1, \ldots, C_n$ are renamed versions of clauses in P such that no pair C_i and C_j of clauses shares common variables. Further, let each C_r, $1 \le r \le n$ be of the form

$$A_r : \rho_r \leftarrow B_1^r : \mu_1^r \& \ldots \& B_{m_r}^r : \mu_{m_r}^r$$

where ρ_r is an annotation term and each μ_k^r is an annotation variable or a constant. Suppose further that $A_1, \ldots, A_n$ are unifiable via an mgu θ and $\rho = \sqcup\{\rho_1, \ldots, \rho_n\}$. Then the clause

$$(A_1 : \rho \leftarrow B_1^1 : \mu_1^1 \& \ldots \& B_{m_1}^1 : \mu_{m_1}^1 \& \ldots \& B_1^n : \mu_1^n \& \ldots \& B_{m_n}^m : \mu_{m_n}^n)\theta$$

is called a *reductant* of P. Here, the expression $\sqcup\{\rho_1, \ldots, \rho_n\}$ is evaluated only if all the ρ_i's are c-annotations; otherwise the complex term-annotation $\sqcup\{\rho_1, \ldots, \rho\}$ becomes the annotation in the head of the reduction.

Theorem 6.18 *If C is a reductant of P, then $P \models C$.*

Proof As C is a reductant of P, it is obtained from clauses $C_1, \ldots, C_n$, where each C_r is a renamed version of a clause in P and no pair C_i and C_j shares common variables. Hence, C is of the form

$$(A_1 : \rho \leftarrow B_1^1 : \mu_1^1 \& \ldots \& B_{m_1}^1 : \mu_{m_1}^1 \& \ldots \& B_1^n : \mu_1^n \& \ldots \& B_{m_n}^n : \mu_{m_n}^n)\theta$$

where $\rho = \sqcup\{\rho_1, \ldots, \rho_n\}$, each C_j, $1 \leq j \leq n$ is of the form

$$A_j : \rho_j \leftarrow B_1^j : \mu_1^j \& \ldots \& B_{m_j}^j : \mu_{m_j}^j$$

and $\{A_1, \ldots, A_n\}$ are unifiable via an mgu θ. Suppose now that I is a model of P, and $C\sigma$ is a strict ground instance of C such that

$$I \models (B_{m_1}^1 : \mu_{m_1}^1 \& \ldots \& B_1^n : \mu_1^n \& \ldots \& B_{m_n}^n : \mu_{m_n}^n)\theta\sigma$$

Hence, I satisfies the body of $C_j\theta\sigma$ for all $1 \leq j \leq i$. I is a model of P and hence a model of each C_j (as the C_j's are only renamed versions of clause in P). Hence, $I(A_1\theta\sigma) \geq \rho_j$ for all $1 \leq j \leq i$, and thus $I(A_1\theta\sigma) \leq \rho = \sqcup\{\rho_1, \ldots, \rho_i\}$.

Proposition 6.1 *Suppose P is a GAP and $A\mu$ is an annotated atom. If $P \models A : \mu$, then there is a reductant of P having the form*

$$A : \rho \leftarrow B_1 : \mu_1 \& \ldots \& B_n : \mu_n$$

such that $\rho \geq \mu$ and $P \models B_1 : \mu_1 \& \ldots \& B_n : \mu_n$.

Proof Suppose $P \models A : \mu$ where A is a ground atom. By Theorem 6.17, $\mu \in T_P \uparrow \omega(A)$ and by Corollary 6.2 there is an integer k such that $\mu \in (T_P \uparrow k)(A)$. We proceed by induction on k.
Base Case: $k = 1$. In this case, $\mu \in (T_P \uparrow 1)(A)$. Hence, there are clauses $C_1, \ldots, C_r, r \geq 1$, in *SGI(P)*, of the form (empty bodies):

$A : \rho \leftarrow$
...
$A : \rho_r \leftarrow$

such that $\mu \leq \rho = \sqcup\{\rho_1, \ldots \rho_r\}$. Since $A : \rho \leftarrow$ is the reductant of the above clause, the base case follows.
Inductive Step: $k = i + 1$. The proof proceeds along the lines of the base case.

Definition 6.36 (*Query*) A *query* is a statement of the form $? - A_1 : \rho_1 \& \ldots \& A_k : \rho_k$, where the $A_i : \rho_i$'s are atoms annotated by a constant or a variable.

We will assume that queries are not necessarily c-annotated, i.e., they may contain annotation variables. If Q is a query $?- A_1 : \rho_1 \& \ldots \& A_k : \rho_k$ then $(\exists)Q$ denote the existential closure of the conjunction of its body literals, $(\exists)(A_1 : \rho_1 \& \ldots \& A_k : \rho_k$.

Definition 6.37 (*Constrained query*) A *constrained query* is a statement of the form:

$$? - A_1 : \rho_1 \& \ldots \& A_k : \rho_k \& Constraint_Q$$

where the $A_1 : \rho_1 \& \ldots \& A_k : \rho_k$ is the *query-part* of Q and $Constraint_Q$ is its *constraint-part*. Here each ρ_j is either a constant from $\mathcal{T}$ or an annotated variable.

A *constrained clause* is of the form:

$A : \psi \leftarrow B_1 : \mu_1 \& \ldots \& B_m : \mu_m \& Constraint_C$

which is a clause in the old sense, augmented by a constraint $Constraint_C$. The notion of satisfaction of such clauses by an interpretation is immediate.

Definition 6.38 Suppose C is a constrained clause $A : \psi \leftarrow B_1 : \mu_1 \& \ldots \& B_m : \mu_m \& Constraint_C$ and Q is a query? $-\ A_1 : \rho_1 \& \ldots \& A_k : \rho_k \& Constraint_Q$ such that

1. C and Q have no (annotation or object) variables in common.
2. A_i and A are unifiable via mgu θ.

Then the *resolvent* of Q and C with respect to A_i is the constrained query Q' below:

$$?-\ (A_1 : \rho_1 \& A_{i-1} : \rho_{i-1} \& B_1 : \mu_1 \& \ldots \& B_m : \mu_m \& A_{i+1} : \rho_{i+1} \& \ldots \& A_k : \rho_k)\theta \& (Constraint_C \& \psi \geq \rho_i \& Constraint_Q) \quad (**)$$

In the above, if θ is not required to be a most general unifier (i.e., θ is allowed to be any unifier), then Q' is called an *unrestricted resolvent* of C and Q with respect to A_i.

A constraint **C** is *solvable* with respect to the semilattice $\mathcal{T}$ and the set $\mathcal{F}$ of interpreted annotation function iff there is an assignment σ of elements in $\mathcal{T}$ to the annotation variables of **C** such that **C** has a solution with respect to $\mathcal{T}$ and $\mathcal{F}$ (**C**σ is evaluated using the intended interpretation of the annotated functions in $\mathcal{F}$).

There is an important class of constraints, called normal constraints.

Definition 6.39 A constraint $(\tau_1 \geq \kappa_1 \& \ldots \& \tau_n \geq \kappa_n)$ is *normal* if

1. Each κ_i is an annotation variable or a constant;
2. If κ_i is a variable, then it does not occur in $\tau_1, \ldots, \tau_i$.

Queries, clauses, and GAPs constrained by normal constraints are called *normal* queries, clauses and GAPs, respectively.

Lemma 6.6 *Suppose $\mathcal{T}$ is a lattice (not necessarily complete). Then,*

1. *If C and Q are a normal clause and a normal query, respectively, then the resolvent of Q and C is a normal query.*
2. *Satisfiability of any normal constraint is decidable.*

Proof (1) Notice that ρ_i in Eq. (**) does not appear in $Constraint_C$ and in ψ, since variables have been renamed before performing the resolution step. Similarly, none of the right-hand side of the inequalities in $Constraint_Q$ appears in ψ.

Therefore, if both $Constraint_C$ and $Constraint_Q$ in the above equation are normal, the constraint in the resolvent, $(Constraint_C \& \psi \geq \rho_i \& Constraint_Q)$, is also normal. Observe that the order of constraints in Eq. (**) above is crucial.
(2) Let **C** be a normal constraint of the form $\tau_1 \geq \kappa_1 \& \ldots \& \tau_n \geq \kappa_n$. Without loss of generality, we assume that the inequalities in **C** with identical variable in the right-hand side are grouped together, i.e., κ_i and κ_j are the same variable, then for all $s, i \leq s \leq j$, κ_s is the same variable as κ_i and κ_j.

This grouping can be achieved by the following re-grouping operation: Suppose **C** has a subsequence of conjuncts $\ldots \& \tau^1 \geq x \& \ldots \& \tau^2 \geq y \& \ldots \& \tau^3 \geq x \ldots$. Because of normality of **C**, x does not appear in τ^2, and hence the whole block of inequalities between $\tau^1 \geq x$ and $\tau^3 \geq x$ can be moved in front of $\tau^2 \geq x$. Clearly, the resulting constraint will still be normal and equivalent to **C**. Repeating this process, we will achieve the desired grouping of conjuncts in **C**.

The test for satisfiablity of **C** in $\mathcal{T}$ now follows:

a. If **C** is an empty constraint, return (**satisfiable**).
b. Let $i_0 \geq 1$ be the maximal integer such that κ_{i_0} is the same symbol as κ_1. Substitute $\top$ for each of the variables occurring in $\tau_1, \ldots, \tau_{i_0}$ (the substitution must be done throughout **C**). Since **C** is normal, none of these variables appears on the right-hand side of **C**. Let the resulting constraint and annotation terms be also denoted by **C** and τ_i's, respectively. Notice that now each of the $\tau_1, \ldots, \tau_{i_0}$ can be evaluated to an element of $\mathcal{T}$.
c. If κ_1 is a constant then

    If $\tau_1 \geq \kappa_1 \& \ldots \& \tau_{i_0} \geq \kappa_{i_0}$ is false in $\mathcal{T}$
    then return (**unsatisfiable**)
    /* The if-condition is verifiable since the τ_i's are ground */

    /* Otherwise */
    Set **C** to $\tau_{i_0+1} \geq \kappa_{i_0+1} \& \ldots \& \tau_n \geq \kappa_n$.
    Rearrange indices of the τ_i's in **C** so that they will start 1. and then go to (a).
d. If κ_1 is a variable then replace it by the greatest lower bound of $\tau_1, \ldots, \tau_{i_0}$, which exists since it was assumed that $\mathcal{T}$ is a lattice. This replacement should be done everywhere in **C**. Then delete the conjuncts $\tau_1 \geq \kappa_1 \& \ldots \& \tau_{i_0} \geq \kappa_{i_0}$ from **C**, as in (c), and go to (a).

Correctness of this algorithm follows immediately from the fact that all functions used in the τ_j in **C** are monotonic. Termination of the algorithm follows from the assumption that all functions in $\mathcal{F}$ are computable.

The above result can be strengthened somewhat by replacing the requirement that $\mathcal{T}$ must be a lattice by a weaker requirement that every finite subset of $\mathcal{T}$ has a (not necessarily greatest) lower bound.

The next result says that if $\mathcal{T}$ is finite then the requirement of normality can be dropped altogether.

Lemma 6.7 *For finite semilattices $\mathcal{T}$, satisfiability of every constraint is decidable.*

Proof Suppose **C** is a constraint over $\mathcal{T}$. Let *GRD* be the set of all instances of this constraint obtained by (uniformly) replacing all occurrences of annotation variables by annotation constants (in particular, constraints in *GRD* are free of annotation variables). As $\mathcal{T}$ is finite, *GRD* is a finite set of ground constraints (since constraints contain no quantifiers, by definition). Now, **C** is solvable iff *some* constraint in *GRD* is solvable. But for ground constraints satisfaction is obviously decidable since they are conjunctions of ground atoms involving decidable predicates only.

Definition 6.40 (*Deduction*) A *deduction* of a constrained query Q_0 from a GAP P is a sequence:

$$Q_0, \langle C_0, \theta_0 \rangle, Q_1, \ldots, Q_n, \langle C_n, \theta_n \rangle, Q_{n+1}$$

such that

1. Q_{i+1} is a resolvent of Q_i via mgu θ; and
2. C_i is a reductant of P that contains no variables in common with Q_i.

When the θ_i's in the above deduction are required to be unifiers but not necessarily mgu's (i.e., the Q_i's, $i \geq 1$, are only required to be unrestricted resolvents), then the above deduction is called an *unrestricted deduction*.

Definition 6.41 (*Refutation*) The deduction $\mathcal{R} = Q_0, \langle C_0, \theta_0 \rangle, Q_1, \ldots, Q_n, \langle C_n, \theta_n \rangle, Q_{n+1}$ of the query Q_0 from P is a *refutation* iff

1. Q_{n+1}, the resolvent of Q_n and C_n, has an empty query-part (i.e., Q_{n+1} is just constraint); and
2. Q_{n+1} is solvable with respect to the lattice $\mathcal{T}$ and the set of annotation functions $\mathcal{F}$.

We will use *SOL*($\mathcal{R}$) to denote the set of solutions of the constraint-part of Q_{n+1}. Unrestricted refutation is defined similarly (where "deduction" by "unrestricted deduction").

The implementation of the above refutation procedure hinges upon two things.

- The ability to solve lattice constraints; and
- The ability to restrict the choice of reductants.

The first is about lattice-theoretical algorithms, and the second is to limit the number of reductants to be considered in refutations.

Definition 6.42 An upper semilattice $\mathcal{T}$ is *n-wide* if for every finite set $E \in \mathcal{T}$, there is a finite subset $E_0 subseteq E$ of at most n elements such that $\sqcup E_0 = \sqcup E$.

An *n-reductant* of a program P is a reductant involving no more than n clauses of P.

It is known that many popular semilattices have finite width. All finite semilattices have this property. Among the infinite ones, the semilattice of the form $[0, 1]^n$ has width n (here $(a_1, \ldots, a_n) \sqcup (b_1, \ldots, b_n) = (a_1 \sqcup b_1, \ldots, a_n \sqcup b_n)$). In particular, $[0, 1]$ and $[0, 1]^2$ are frequently used in expert systems.

To show that $[0, 1]^2$ is 2-wide, let $\alpha_1 = [a_1, b_1], \ldots, \alpha_k = [a_k, b_k]$ be a finite set of pairs of real numbers in the interval $[0, 1]$. Let a_i (b_j) be the maximal element among the $a_1, \ldots, a_k$ $(b_1, \ldots, b_k)$. Then, $\alpha_i \sqcup \alpha_j = \sqcup\{\alpha_1, \ldots, \alpha_k\}$, which proves that $[0, 1]^2$ is 2-wide.

If $\mathscr{T}$ is n-wide, then in building refutations it suffices to consider n-reductants only. This limits the choice of clauses to solve with to a *finite* set of n-reductants.

Theorem 6.19 (Soundness) *Suppose P is a GAP and Q is a constrained query such that*

$$Q_0, \langle C_0, \theta_0 \rangle, Q_1, \ldots, Q_n, \langle C_n, \theta_n \rangle, Q_{n+1}$$

is a refutation of Q_0 from the GAP P. Let σ be any solution for the constraint-part of Q_{n+1}. Then, $Q_0\sigma$ is an annotation-variable-free query obtained by replacing all annotation variables in Q_0 by the annotation constants specified in σ. We claim that:

$$P \models (\forall)(Q_0\sigma)\theta_0\theta_1 \ldots \theta_n.$$

Here, $(\forall)Q_0$ denotes a conjunction of body literals of query Q_0 universally quantified.

Proof We proceed by induction on n, the length of the refutation of Q_0 from P. Base case: $n = 1$. Then, Q_0 contains exactly one annotation atom, denoted $A : \mu$. Hence, C_0 is of the form

$$D_0 : \rho_0 \leftarrow$$

such that $A\theta_0 = D\theta_0$ and the constraint $(\rho_0 > \mu)$ is solvable. Let σ be any solution of this constraint and let I be a model of P. Then, I is a model of C_0 (as P entails all its reductants) and, in particular,

$$I \models (\forall)(D_0 : \rho)$$

where ρ is obtained from ρ_0 by instantiating all annotation variables to $\top$, the top element of $\mathscr{T}$. In particular,

$$I \models (\forall)A_0\theta : \rho$$

and as $\rho = \rho_0\sigma \geq \mu\sigma$ (due to the monotonicity of annotation functions in $\mathscr{F}$), it follows that

$$I \models (\forall)A_0\theta_0 : \rho$$

Inductive step: $n > 1$. In this case,

$$Q_1, \langle C_1, \theta_1 \rangle, Q_2, \ldots, Q_n, \langle C_n, \theta_n \rangle, Q_{n+1}$$

is a refutation of Q_1, where the body of Q_{n+1} is a pure constraint. Let σ be any solution of the constraint-part of Q_{n+1}. Suppose Q_0 is

$$?-\ A_1 : \mu_1 \& \ldots \& A_k : \mu_k \& Constraint_{Q_0}.$$

If C_0 is of the form

$$A : \rho \leftarrow B_1 : \rho_1 \& \ldots \& B_r : \rho_r$$

and $A\theta_0 = A_i\theta_0$ (i.e., Q_0 and C_0 resolve on atom $A_i : \rho_i$, then Q_1 is of the form:

$$?-\ A_1 : \mu_1 \& \ldots \& A_{i-1} : \mu_{i-1} \& B_1 : \rho_1 \& \ldots \& B_r : \rho_r \& A_{i+1} : \mu_{i+1} \& \ldots \& A_n : \mu_n \& \ (Constraint_{Q_0} \& (\rho \geq \mu_i).$$

By the inductive hypothesis, we may assume that

$$P \models (\forall)(Q_1\sigma\theta_1 \ldots \theta_n).$$

Suppose now that I is a model of P. Then, as

$$I \models (\forall)(Q_1\sigma\theta_1 \ldots \theta_n), \qquad (+)$$

it follows that

$$I \models (\forall)(B_1 : \rho_1\sigma \& \ldots \& B_r : \rho_r\sigma)\theta_1 \ldots \theta_n).$$

Hence, since $I \models C_0$ (as C_0 is a reductant of P):

$$I \models (\forall)A\theta_1 \ldots \theta_n : \rho\sigma.$$

As $A\theta_0 = A_i\theta_0$, we conclude that:

$$I \models (\forall)A_i\theta_0 \ldots \theta_n : \rho\sigma.$$

Recall that the constraint-part of Q_1 (and hence of Q_{n+1}) contains the constraint $\rho \geq \mu_i$. As σ is a solution of the constraint Q_{n+1}, it follows that $\rho\sigma \geq \mu_i\sigma$ holds true in $\mathcal{T}$. Thus,

$$I \models (\forall)A_i\theta_0 \ldots \theta_n : \mu_i.$$

Finally, this and (+) imply

$$I \models (\forall)Q_0\sigma\theta_0 \ldots \theta_n.$$

Lemma 6.8 (Mgu lemma) *Suppose P is a GAP and Q is a query. Suppose there is an unrestricted refutation $\mathcal{R}$ of Q such that $\sigma \in$ SOL($\mathcal{R}$). Then, there is a refutation $\mathcal{R}'$ of Q such that $\sigma \in$ SOL($\mathcal{R}'$).*

Proof Similar to the proof of the mgu lemma in classical logic programming (Lloyd [111]).

Lemma 6.9 (Lifting lemma) *Suppose P is a GAP and Q is a normal query. Suppose σ is an assignment of c-annotations to some (not necessarily all) annotation variables in Q and let θ be a substitution for object variables. If there is a refutation $\mathcal{R}$ of $Q\sigma\theta$ from P, then there is a refutation $\mathcal{R}'$ of Q from P.*

Proof Similar to the proof of the lifting lemma in classical logic programming (Lloyd [111]).

Theorem 6.20 (Completeness theorem) *Suppose that $\mathcal{T}$ is a lattice, P is a GAP and Q is a normal query. Suppose $P \models (\exists)Q$. Then, there is a refutation of Q from P. Moreover, if $\mathcal{T}$ is n-wide then Q can be refuted solely using n-reductants of P.*

Proof Suppose Q is

$$?- \ A_1 : \mu_1 \& \ldots \& A_k : \mu_k \& \mathbf{C}_Q.$$

where $\mathbf{C}_Q$ is the constraint-part of Q. As $P \models (\exists)Q$, it follows from Theorem 6.17 that $T_P \uparrow \omega \models (\exists)Q$ and hence, there is an integer n such that $T_P \uparrow n \models (\exists)Q$. We first proceed by induction on n to show that there is an *unrestricted* refutation of Q from P.
Base case: $m = 1$. In this case, $k = 1$ and there is a reductant C of P of the form:

$$A_1 : \rho \leftarrow$$

such that the constraint $C_1 \equiv (\mathbf{C}_Q \& \rho \geq \mu_1)$ is solvable. Hence, $q, \langle C, \theta \rangle, Q_1$, where Q_1 is the goal?– C_1, is an unrestricted refutation of Q from P.
Inductive step: $m = n + 1$. Suppose now that $T_P \uparrow (n + 1) = (\exists)Q$. In particular, there is a variable-free instance $Q\sigma\theta$ of Q (here, σ is an assignment of c-annotation to annotation variable and θ is a ground substitution for object variables) such that $T_P \uparrow (n+1) \models Q\sigma\theta$. By the definition of T_P (Definition 6.33), this implies that for each $1 \leq i \leq k$, there is a reductant of P, denoted C_i, having a ground instance of the form

$$A_i : \rho_i \leftarrow B_1^i : \psi_1^i \& \ldots \& B_{r_i}^i : \psi_{r_i}^i$$

such that $T_p \uparrow n \models (B_1^i : \psi_1^i \& \ldots \& B_{r_i}^i : \psi_{r_i}^i)$ and $(\rho_1 \geq \mu_1 \& \ldots \& \rho_k \geq \mu_k \& \mathbf{C}_Q)$ is solvable.

Furthermore, if $\mathcal{T}$ is n-wide, then we can choose each C_i above to be a n-reductant. Indeed, in Definition 6.33, in order to obtain $T_P(I)(A)$ one needs to take all possible finite least upper bounds of the elements of the set

$$\{f(\mu_1, \ldots, \mu_n \mid A : f(\mu_1, \ldots, \mu_n \leftarrow B_1 : \mu_1 \& \ldots \& B_n : \mu_n \text{ is in } SGI(P),$$
$$\text{and } I \models (B_1 : \mu_1 \& \ldots \& B_n : \mu_n)\}$$

which amounts to taking all possible reductants of $SGI(P)$. However, if $\mathcal{T}$ is n-wide, one only needs to take the least upper bounds of up to n elements of the above set, which amounts to taking n-reductants only.

As $\mathbf{C}_Q$ is normal, and as annotation variables are renamed prior to resolution, $\rho_1 \geq \mu_1 \& \ldots \& \rho_k \geq \mu_k \& \mathbf{C}_Q$ is a normal query. Hence, for all $1 \leq i \leq k$ and for all $1 \leq j \leq r_i$, we may assume, by the induction hypothesis, that there is an unrestricted refutation, denoted $\mathcal{R}_j^i$ of $B_j^i : \psi_j^i$. Then, the k-resolution steps that involve Q and the clauses $C_1, \ldots, C_k$ above, followed by

$$\mathcal{R} = \mathcal{R}_1^1 \mathcal{R}_1^2 \ldots, \mathcal{R}_{r_1}^1 \ldots \mathcal{R}_1^k \mathcal{R}_2^k \ldots \mathcal{R}_{r_k}^k$$

is an unrestricted refutation $Q\sigma\theta$ from P. By the Lifting Lemma, there is an unrestricted refutation of Q from P.

Thus, we know that there is an unrestricted refutation of Q from P. By the Mgu Lemma, it now follows that there is a refutation of Q from P.

It should be noted that there is no similar completeness result for r-entailment, even in the case of acceptable GAPs.

Kifer and Subrahmanian further discussed multiple-valued logics, bilattice-valued logics and temporal reasoning as applications of GAPs. These topics obviously defend the usefulness of GAPs. GAPs can interpret Fitting's bilattice-based logic programming.

The syntax of a bilattice logic program in Fitting's formulation is similar to that of an ordinary logic program, except that the body of a clause may be an arbitrary first-order formula constructed out of $\wedge_t$, $\vee_t \wedge_k$, $\vee_k$ and $\neg$. Negation is interpreted as a unary function on truth-values such that $\mu_1 \leq_k \mu_2$ iff $\neg\mu_1 \vee_k \neg\mu_2$ and $\mu_1 \vee_t \mu_2$ iff $\mu_2 \vee_t \neg\mu_1$.

Although annotated clauses have a two-valued satisfaction relation, formulas in bilattice logics may assume any truth-value from $\mathcal{R}$. The truth order is to allow defining the logical connectives $\wedge_t \vee_t$ and $\neg$ without having to bother with specifics of the set of truth-values $\mathcal{R}$. However, the truth order plays no role in the semantics of annotated logics.

The role of the knowledge order is to give meaning to logical implication, and it is used in a similar way by both annotated and bilattice logics. This means that Fitting's theory of bilattice based logic programming uses the knowledge order in a more essential way that the truth order.

In fact, Fitting's interpretations are the same as r-interpretations of GAPs. In other words, they are functions from the Herbrand base of P to $\mathcal{R}$. These functions are extended to arbitrary formulas by distributing them through the connectives $\wedge_t$, $\vee_t$, $\wedge_k$, $\vee_k$ and $\neg$. Associated with a program P is an operation V_P that maps interpretations to interpretations as follows:

$V_P(I)(A) = \sqcup\{\mu \mid A \leftarrow L_1\&\ldots\&L_n$ is a ground instance of a clause in P and $\mu = I(L_1\&\ldots\&L_n)\}$

Given a bilattice-based logic program, we can translate it into a GAP, denoted $\mathbf{bl}(P)$. Let C denote a clause $A \leftarrow Body(L_1, \ldots, L_n)$ in P, where $Body(L_1, \ldots, L_n)$ is Fitting's formula involving atomic literals $L_1, \ldots, L_n$. Then, the corresponding clause $\mathbf{bl}(C)$ has the form:

$$A : f_{Body(L_1,\ldots,L_n)}(T_1, \ldots, T_n) \leftarrow L_1 : T_1\&\ldots\&L_n : T_n$$

where $T_1, \ldots,_n$ are annotation variables and the function $f_{Body(L_1,\ldots,L_n)}$ is defined as follows:

1. If $Body(L_1, \ldots, L_n) = Body_1(L_1, \ldots, L_k) \vee_t Body_2(L_{k+1}, \ldots, L_n)$ then $f_{Body}(T_1, \ldots, T_n) = f_{Body_1}(T_1, \ldots, T_k) \sqcup_t f_{Body_2}(T_{k+1}, \ldots, T_n)$
2. If $Body(L_1, \ldots, L_n) = Body_1(L_1, \ldots, L_k) \vee_k Body_2(L_{k+1}, \ldots, L_n)$ then $f_{Body}(T_1, \ldots, T_n) = f_{Body_1}(T_1, \ldots, T_k) \sqcup_k f_{Body_2}(T_{k+1}, \ldots, T_n)$

3. If $Body(L_1, \ldots, L_n) = Body_1(L_1, \ldots, L_k) \wedge_t Body_2(L_{k+1}, \ldots, L_n)$ then $f_{Body}(T_1, \ldots, T_n) = f_{Body_1}(T_1, \ldots, T_k) \sqcap_t f_{Body_2}(T_{k+1}, \ldots, T_n)$
4. If $Body(L_1, \ldots, L_n) = Body_1(L_1, \ldots, L_k) \wedge_k Body_2(L_{k+1}, \ldots, L_n)$ then $f_{Body}(T_1, \ldots, T_n) = f_{Body_1}(T_1, \ldots, T_k) \sqcap_k f_{Body_2}(T_{k+1}, \ldots, T_n)$
5. If $Body(L_1, \ldots, L_n) = \neg Body_1(L_1, \ldots, L_n)$ then $\neg f_{Body}(T_1, \ldots, T_n) = f_{Body_1}(T_1, \ldots, T_k)$

Kifer and Subrahmanian proved that for a bilattice-based logic program P, $V_P = R_{\mathbf{bl}(P)}$ and that the models of a billatice logic program P are also the models of the GAP $\mathbf{bl}(P)$ and their least models coincide.

Thus, billatice-based logics can be interpreted as annotated logics. However, the full logics for GAPs should be worked out. It seems also that there are no implementations of GAPs.

Logic programming languages based on annotated logics were studied by other people. Abe and his group implemented a paraconsistent logic programming language *Paralog*; see da Costa et al. [68]. Later, they proposed a paraconsistent evidential logic programming language *ParaLog-e* as a variant of Paralog; see Avila et al. [41] for details.

Nakamatsu and Suzuki [129, 130] proposed *annotated logic program with strong negation* (ALPSN) to deal with inconsistency and non-monotonic reasoning. Later, as a generalization of the framework, Nakamatsu [125] proposed *vector annotated logic program with strong negation* (VALPSN) and their extensions.

Lu [114] proposed *signed formula logic programming* which is closely related to annotated logic programming. The sign in signed formula logic programming is similar to the annotation in GAPs, but the sign is a set of truth-values without any ordering. In this regard, signed formula logic programming is a simplified version of GAPs, and it can be also applied to paraconsistent reasoning. Lu also developed its semantics and a signed resolution as a query procedure with soundness and completeness results; also see Lu et al. [115].

6.3 Knowledge Representation

Knowledge representation is to describe human knowledge in a reasonable form. Since human knowledge is very complex, we face difficulties with the subject. Knowledge representation has been studied in AI, and some formalisms are closely related to *object-oriented programming*.

In computer science including AI, a good solution for a given problem depends on a good representation. For most AI applications, the choice of knowledge representation formalism is even more difficult, since the criteria for such choices are less clear.

Though no general consensus exists of what knowledge representation is, many schemes have been proposed to represent and store knowledge. Many such schemes have been successfully used as a foundation for the implementation of some existing systems. There are, however, several characteristics of knowledge that are not yet well understood, such as defaults and inconsistencies. Until a better comprehension of such characteristics is achieved, the representation of knowledge will remain as an active field of study.

There are several schemes to represent knowledge. Two schemes that better *capture* the knowledge concerning objects and their properties are *semantic network* and *frame*. The latter can be seen as a generalization of the former. A frame system was proposed by Minsky [120] who criticized classical logic as a knowledge representation language. In fact, a frame can naturally capture inheritance of knowledge and *non-monotonic reasoning*. Inheritance naturally leads to non-monotonic reasoning, in which new information invalidates the old conclusions.

These features defend frame systems, since they cannot be formalized logic-based formalisms. Since the 1980s, AI researchers have investigated the so-called *non-monotonic logic* which is a logical system capable of formalizing non-monotonic reasoning. There are several non-monotonic logics, which include the non-monotonic logic of McDermott and Doyle [118, 119], the *autoepistemic logic* of Moore [121] and the *default logic* of Reiter [140].

However, none of the existing non-monotonic formalisms can adequately deal with issues related to inconsistency. It is not appropriate for formalizing non-monotonic logics. The difficulty lies in the fact the most non-monotonic logics are based on classical logic although they try to modify or extend it.

To implement knowledge representation systems like frame systems handling inconsistencies with logical foundations, we need paraconsistent logics. In this regard, annotated logics are promising. Annotated logics can serve as foundations for knowledge representation and object-oriented programming.

One of the notable systems in this direction is *F-logic* (Frame Logic) due to Kifer et al. [97]. F-logic can deal with most of the structural aspects of object-oriented and frame-based languages, which include object identity, complex objects, and inheritance. In addition, F-logic has a model-theoretic semantics and a sound and complete resolution-based proof theory.

Now, we review F-logic in detail. *Object-oriented* is an approach to computer science including programming and databases in which an *object* is regarded as a first-class citizen. Various object-oriented approaches have been studied, and we now have many successful applications. However, most approaches have no theoretical foundations.

Frame-based languages like the frame system discussed above share many features of object-oriented languages, and they face some difficulties. F-logic provides a unified framework both for frame and object-oriented systems inspired by the work on annotated logics. F-logic does not have all the features of annotated logics, in particular, the representation of inconsistent knowledge, but the basic ideas are essentially similar.

The alphabet of F-logic language $\mathscr{L}$ consists of the following:

- a set of *object constructors* $\mathscr{F}$
- an infinite set of *variables* $\mathscr{V}$

- logical connective and quantifiers; $\wedge, \vee, \neg, \forall, \exists$
- auxiliary symbols, such as, (,), [,], etc.

Object constructors (elements of $\mathscr{F}$) play the role of function symbols of F-logic. Each function symbols has an arity, which is a non-negative integer that determines the number of arguments this symbol can take. Symbols of arity 0 are also called *constants*; symbols of arity ≥ 1 are used to construct larger terms out of simpler ones.

An *id-term* is a usual first-order term composed of function symbols and variables, as in predicate calculus. The set of all ground id-terms is denoted by $U(\mathscr{F})$, which is also known as the *Herbrand universe*.

A language of F-logic consists of a set of formulas consists of a set of formulas constructed out of the alphabet symbols. As in many other logics, formulas are built out of simpler formulas using usual logical connectives and quantifiers. The simplest kind of formulas are called *molecular formulas*.

Definition 6.43 (*Molecular formulas*) A *molecular formula* in F-logic is one of the following statements:

1. An *is-a assertion* of the form $C::D$ or of the form $O::C$, where C, D, and O are id-terms.
2. An *object molecule* of the form O[a';'-separated list of *method expressions*].
   A *method expression* can be either a *non-inheritable data expression*, an *inheritable data expression*, or a *signature expression*.
   - Non-inheritable data expressions take one of the following two forms:
     – A non-inheritable *scalar* expression ($k \geq 0$):
       $ScalarMethod@Q_1, \ldots, Q_k \rightarrow T$
     – A non-inheritable *set-valued* expression ($l, m \geq 0$):
       $SetMethod@R_1, \ldots, R_l \twoheadrightarrow \{S_1, \ldots, S_m\}$
   - *Inheritable* scalar and set-valued data expressions are like non-inheritable expressions except that "$\rightarrow$" is replaced with "$\bullet\!\!\rightarrow$" and "$\twoheadrightarrow$" is replaced with "$\bullet\!\!\twoheadrightarrow$".
   - *Signature expressions* also take two forms:
     – A *scalar* signature expression ($n, r \geq 0$):
       $ScalarMethod@V_1, \ldots, V_n \Rightarrow (A_1, \ldots, A_n)$
     – A *set-valued* signature expression ($s, t \geq 0$):
       $SetMethod@W_1, \ldots, W_s \Rrightarrow (B_1, \ldots, B_n)$

The first is-a assertion in (1), $C::D$, states that C is a *nonstrict* subclass of D (i.e., inclusive of the case when C and D denote the same class). The second assertion, $O : C$, states that O is a member of class C.

In (2), O is an id-term that denotes an object. *ScalarMethod* and *SetMethod* are also id-terms. However, the syntactic context of *ScalarMethod* indicates that it is invoked on O as a scalar method, while the content of *SetMethod* indicates a set-valued invocation.

Double-headed arrows $\twoheadrightarrow$, $\bullet\!\!\twoheadrightarrow$, $\Rightarrow\!\!\!\!\Rightarrow$ indicate that *SetMethod* denotes a set-valued function. The single-headed arrows, $\rightarrow$, $\bullet\!\!\rightarrow$, $\Rightarrow$, indicate that the corresponding method is scalar.

In the above data expressions, T and S_i are id-terms that represent output of the respective methods, *ScalarMethod* and *SetMethod*, when they are invoked on the host-object O with the arguments $Q_1, \ldots, Q_k$ and $R_1, \ldots, R_k$, respectively. The arguments are id-terms.

In the signature expressions, A_i and B_i are id-terms that represent *types* of the results returned by the respective methods when they are invoked on an object of class C with arguments of types $V_1, \ldots, V_n$ and $W_1, \ldots, W_n$, respectively; these arguments are also id-terms.

The notation "(...)" in signature expressions is intended to say that the output of the method must belong to *all* the classes listed in parentheses to the right of "$\Rightarrow$" and "$\Rightarrow\!\!\!\!\Rightarrow$".

Definition 6.44 (*F-formulas*) *F-formulas* are built out of simpler F-formulas by means of logical connectives and quantifiers.

1. Molecular formulas are F-formulas;
2. $\varphi \wedge \psi, \varphi \vee \psi, \neg\varphi$ are F-formulas, if so are φ and ψ;
3. $\forall X \varphi, \exists Y \psi$ are formulas, if so are φ, ψ and X, Y are variables.

In addition, we define a *literal* to be either a molecular formula or its negation.

Next, we describe a semantics for F-logic. Given a pair of sets, U and V, we use *Total*(U, V) to denote the set of all total functions $U \rightarrow V$; similarly *Partial*(U, V) stands for the set of all partial functions $U \rightarrow V$. The powerset of U is denoted by $\wp(U)$.

Given a collection of sets, $\{S_i\}_{i \in N}$ parameterized by natural numbers, $\prod_{i=1}^{\infty} S_i$ denotes the Cartesian product of the S_i's, that is, the set of all infinite tuples $\langle s_1, \ldots, s_n, \ldots \rangle$.

Given an F-language $\mathscr{L}$, an *F-structure* is expressed as a tuple

$$\mathbf{I} = \langle U, \preceq_U, \in_U, I_{\mathscr{F}}, I_{\rightarrow}, I_{\twoheadrightarrow}, I_{\bullet\!\rightarrow}, I_{\bullet\!\twoheadrightarrow}, I_{\Rightarrow}, I_{\Rightarrow\!\!\!\Rightarrow} \rangle.$$

Here, U is the domain of $\mathbf{I}$, $\preceq_U$ is an irreflexive partial order on U, and $\in_U$ is a binary relation. The ordering $\preceq_U$ on U is a semantic counterpart of the subclass-relationship, i.e., $a \preceq_U b$ is interpreted as a statement that a is a subclass of b.

The binary relation $\in_U$ is used to model class membership, i.e., $a \in_U b$ is interpreted as a statement that a is a member of class b. The two binary relationships, $\preceq_U$ and $\in_U$, are related as follows: if $a \in_U b$ and $b \preceq_U c$ then $a \in_U c$.

We can view U as a set of all *actual* objects in a *possible world* $\mathbf{I}$. Ground id-terms (the element of $U(\mathscr{F})$) play the role of logical object id's. They are interpreted by the object in U via the mapping $I_{\mathscr{F}} \rightarrow \bigcup_{i=0}^{\infty} Total(U^i, U)$. This mapping interprets each k-ary object constructor, $f \in \mathcal{F}$ by a function $U^k \rightarrow U$. For $k = 0, I_{\mathcal{F}}(f)$ can be identified with an element of U.

The remaining six symbols in **I** denote mappings for interpreting each of the six types of method expressions in F-logic.

$$I_{\rightarrow}, I_{\bullet\rightarrow} : U \rightarrow \prod_{k=0}^{\infty} \mathit{Partial}(U^{k+1}, U)$$

Each of these mappings associates a tuple of partial functions $\{f_k \mid U^{k+1} \rightarrow U, k \geq 0\}$ with every element of U; there is exactly one such f_k in the tuple, for every method-ary $k \geq 0$. In other words, the same method can be invoked with different arities.

$$I_{\rightarrow\!\!\rightarrow}, I_{\bullet\rightarrow\!\!\rightarrow} : U \rightarrow \prod_{k=0}^{\infty} \mathit{Partial}(U^{k+1}, \wp(U))$$

For every method-arity k, each of these mappings associates a partial function $\{U^{k+1} \rightarrow \wp(U)\}$ with each element of U. Note that each element of U has four different sets of interpretations: two provided by $I_{\rightarrow}$ and $I_{\bullet\rightarrow}$ and two provided by $I_{\rightarrow\!\!\rightarrow}$ and $I_{\bullet\rightarrow\!\!\rightarrow}$.

$$L_{\Rightarrow} : U \rightarrow \prod_{i=0}^{\infty} \mathit{PartialAntiMonotone}_{\leq_U}(U^{i+1}, \wp_{\uparrow}(U))$$
$$L_{\Rightarrow\!\!\!\!\!\Rightarrow} : U \rightarrow \prod_{i=0}^{\infty} \mathit{PartialAntiMonotone}_{\leq_U}(U^{i+1}, \wp_{\uparrow}(U))$$

Here, $\wp_{\uparrow}(U)$ is a set of all *upward-closed* subsets of U. A set $V \subseteq U$ is *upward closed* if $v \in V$ and $v \leq_U v'$ imply $v' \in V$. $\mathit{PartialAntiMonotone}_{\leq_u}(U^{i+1}, \wp_{\uparrow}(U))$ denotes the set of partial *anti-monotonic* function from U^{i+1} to $\wp_{\uparrow}(U)$.

For a partial function $\rho : U^k \rightarrow \wp_{\uparrow}(U)$, *anti-monotonicity* means that if $\overrightarrow{u}, \overrightarrow{v} \in U^k$, $\overrightarrow{v} \leq_U \overrightarrow{u}$ and $\rho(u)$ is defined, then $\rho(\overrightarrow{v})$ is also defined and $\rho(\overrightarrow{v}) \supseteq \rho(\overrightarrow{u})$.

A *variable assignment*, v, is a mapping from the set of variables, $\mathcal{V}$, to the domain U. variable assignments extend to id-terms in the usual way] $v(d) = \mathbf{I}_{\mathcal{F}}(d)$ if $d \in \mathcal{F}$ has arity 0 and recursively, $v(f(\ldots, T, \ldots)) = \mathbf{F}_{\mathcal{F}}(f)(\ldots, v(T), \ldots)$.

Let $bf I$ be an F-structure and v a variable assignment. Intuitively T[..., a molecule is true under **I** with respect to a variable assignment v, denoted $\mathbf{I} \models T[\ldots]$, if the object $v(T)$ in **I** has properties that the formula $T[\ldots]$ says it has. An is-a molecule, $P::Q$ or $P : Q$ i true if the objects involved, $v(P)$ and $v(Q)$, are related via $\leq_U$ or $\in_U$ to each other.

Definition 6.45 (*Satisfaction of F-molecules*) Let **I** be an F-structure and G be an F-molecule. We write $\mathbf{I} \models_v G$ iff all the following holds]

1. When G is an is-a assertion then:

   (i) $v(Q) \leq_U v(P)$, if $G = Q::P$; or
   $v(Q) \in_U v(P)$ if $G : P$.

2. When G is an object molecule of the form $O[a$‘;’*−separated list of method expressions*] then for every method expression E in G, the following conditions must hold:

(ii) If E is a non-inheritable scalar data expressions of the form $ScalM@Q_1, \ldots, Q_k \rightarrow T$, the element $I^{(k)}_{\rightarrow}(v(ScalM))(v(O), v(Q_1), \ldots, v(Q_k))$ must be defined and equal $v(T)$.
Similar conditions must hold if E is an inheritable scalar data expression, except that $I^{(k)}_{\rightarrow}$ should be replaced with $I^{(k)}_{\bullet\rightarrow}$.

(iii) If E is a non-inheritable set-valued data expression, of the form $SetM@R_1, \ldots, R_l \twoheadrightarrow \{S_1, \ldots, S_m\}$, the set $\mathbf{I}^{(l)}_{\twoheadrightarrow}(v(SetM))(v(O), v(R_1), \ldots, v(R_l))$ must be defined and contain the set $\{v(S_1), \ldots, v(S_m)\}$.
Similarly conditions must hold if E is an inheritable set-valued data expression, except that $I^{k}_{\twoheadrightarrow}$ should be replaced with $I^{(k)}_{\bullet\twoheadrightarrow}$.

(iv) If E is a scalar signature expression, of the form $ScalM@Q_1, \ldots, Q_n \Rightarrow (R_1, \ldots, R_u$, then the set $\mathbf{I}^{(n)}_{\Rightarrow}(v(Scal(M))(v(O), v(Q_1), \ldots, v(Q_n))$ must be defined and contain $\{v(R_1), \ldots, v(R_n)\}$.

(v) If e is a set-valued signature expression, of the form $SetM@V_1, \ldots, V_s \Rightarrow\!\!\!\!\Rightarrow (W_1, \ldots, W_v)$, the set $\mathbf{I}^{(s)}_{\twoheadrightarrow}(v(SetM))(v(O), v(V_1), \ldots, v(V_s))$ must be defined and contain $\{s(W_1), \ldots, s(W_v)\}$.

Here, (i) says that the object $v(Q)$ must be a subclass or a member of the class $v(P)$.

Conditions (ii) and (iii) say that in case of a data expression the interpreting function must be defined using appropriate arguments and yield results compatible with those specified by the expression.

Conditions (iv) and (v) say that, for a signature expression, the type of a method (*ScalM* or *SetM*) specified by the expression must comply with the type assigned to this method by **I**.

The meaning of the formulas $\varphi \wedge \psi$, $\varphi \vee \psi$ and $\neg\varphi$ and the meanings of quantifiers are defined in the standard way:

$\mathbf{I} \models_v \varphi \wedge \psi$ iff $\mathbf{I} \models_v \varphi$ and $\mathbf{I} \models_v \psi$
$\mathbf{I} \models_v \varphi \vee \psi$ iff $\mathbf{I} \models_v \varphi$ or $\mathbf{I} \models_v \psi$
$\mathbf{I} \models_v \neg\varphi$ iff $\mathbf{I} \not\models \varphi$
$\mathbf{I} \models_v (\forall X)\varphi$ iff $\mathbf{I} \models_\mu$ for every μ that agrees with v everywhere except possibly on X.
$\mathbf{I} \models_v (\exists X)\varphi$ iff $\mathbf{I} \models_\mu$ for some μ that agrees with v everywhere except possibly on X.

For a closed formula φ, we can omit the meaning of v and simply write $\mathbf{I} \models \varphi$, since the meaning of a closed formula is independent of the choice of variable assignments.

An F-structure **I** is a *model* of a closed formulas ψ iff $\mathbf{I} \models \psi$. If **S** is a set of formulas and φ is a formula, we write $\mathbf{S} \models \psi$ (φ is *entailed* by **S**) iff φ is true i every model of **S**.

Predicates in first-order logic can be encoded as F-molecules. To encode n-ary predicate symbol p, we introduce a new class for which we will conveniently reuse the same symbol p.

Let *p-tuple* be a new n-ary function symbol. We then assert $(\forall X_1 \ldots \forall X_n)(p\text{-}tuple(X_1, \ldots, X_n) : p$ and write classical atoms of the form $p(T_1, \ldots, T_n)$ as molecules of the form:

$p\text{-}tuple(T_1, \ldots, T_n)[arg_1 \rightarrow T_1; \ldots; arg_n \rightarrow T_n]$

which asserts that there is a p-relationship among $T_1, \ldots, T_n$.

To incorporate predicates directly, we can extend the notion of F-language with a new set $\mathscr{P}$ of *predicate symbols*. If $p \in \mathscr{P}$ is an n-ary predicate symbol and $T, \ldots, T_n$ are id-terms, then $p(T_1, \ldots, T_n)$ is a *predicate molecule* (P-molecule).

A (generalized) *molecule formula* is now either an F-molecule or P-molecule. A *literal* is either a molecule formula or negated molecule formula.

Predicate symbols are interpreted as relations on U using the function $I_{\mathscr{P}}$ which become parts of the definition of the F-structure:

$I_{\mathscr{P}}(p) \subseteq U^n$ for any n-ary predicate symbol $p \in \mathscr{P}$

Given a F-structure $\mathbf{I} = \langle U, \preceq_U, \in_U, \mathbf{I}_{\mathscr{F}}, \mathbf{I}_{\mathscr{P}}, \mathbf{I}_{\rightarrow}, \mathbf{I}_{\twoheadrightarrow}, \mathbf{I}_{\bullet\rightarrow}, \mathbf{I}_{\bullet\twoheadrightarrow}, \mathbf{I}_{\Rightarrow}, \mathbf{I}_{\Rrightarrow} \rangle$ and a variable assignment v, we write $\mathbf{I} \models_v p(T_1, \ldots, T_n)$ if

$\langle v(T_1), \ldots, v(T_n) \rangle \in \mathbf{I}_{\mathscr{P}}(p)$.

We also fix a diagonal interpretation for the equality predicate:

$\mathbf{I}_{\mathscr{P}}(\doteq) =_{def} \{\langle a, a \rangle \mid a \in U\}$

This implies that if T and S are id-terms, then $\mathbf{I} \models_v T \doteq S$ iff $v(T) = v(S)$.

The equality $\doteq$ satisfies the following properties:

Reflexivity
For all $p \in U(\mathscr{F})$, $\mathbf{I} \models p \doteq p$
Symmetry
If $\mathbf{I} \models p \doteq q$ then $\mathbf{I} \models q \doteq p$
Transitivity
If $\mathbf{I} \models p \doteq q$ and $\mathbf{I} \models q \doteq r$ then $\mathbf{I} \models p \doteq r$
Substitution
If $\mathbf{I} \models s \doteq t \wedge L$, and L' is obtained by replacing an occurrence of s in L with t, then $\mathbf{I} \models L'$

The is-a relation :: satisfies the following properties:

IS-A Reflexivity
$\mathbf{I} \models p{::}p$
IS-A transitivity
If $\mathbf{I} \models p{::}q$ and $\mathbf{I} \models q{::}r$ then $\mathbf{I} \models p{::}r$
IS-A acyclicity
If $\mathbf{I} \models p{::}q$ and $\mathbf{I} \models q{::}p$ then $\mathbf{I} \models p \doteq q$
Subclass inclusion
If $\mathbf{I} \models p{::}q$ and $\mathbf{I} \models q{::}r$ then $\mathbf{I} \models p{::}r$

We turn to a resolution-based proof theory for F-logic which uses the prenex normal form. All formulas are converted into the prenex normal form and then skolemized. Skolemized formulas are converted into an equivalent *clausal form*.

Skolemization in F-logic is analogous to the classical case, since id-terms are identical to terms in predicate logic and quantification is defined similarly.

Theorem 6.21 (Skolem's theorem) *Let φ be an F-formula and φ' be its Skolemization. Then, φ is unsatisfiable iff so is φ'.*

Given an F-language $\mathcal{L}$ with $\mathcal{F}$ as its sets of function symbols and $\cap P$ as the set of predicate symbols, the *Herbrand universe* of $\mathcal{L}$ is $U(\mathcal{F})$, the set of all ground id-terms. The *Herbrand base* of $\mathcal{L}$, $\mathcal{HB}(\mathcal{L})$, is the set of all ground molecules (including P-molecules and equality).

Let $\mathbf{H}$ be a subset of $\mathcal{HB}(\mathcal{L})$; it is a *Herbrand structure* (H-structure) of $\mathcal{L}$ if it is closed under the logical implication $\models$. From the definition, closure properties above are obtained by replacing $\mathbf{I} \models \varphi$ with a statement of the form $\varphi \in \mathbf{H}$.

Definition 6.46 (*Satisfaction of Formulas by H-structures*) Let $\mathbf{H}$ be a $\mathbf{H}$-structure. Then, we define the following:

- A ground molecule t is true in $\mathbf{H}$, denoted $\mathbf{H} \models t$ iff $t \in \mathbf{H}$;
- A ground negative literal $\neg t$ is true in $\mathbf{H}$, denoted $\mathbf{H} \models \neg t$ iff $t \notin \mathbf{H}$;
- A ground clause $L_1 \vee \ldots \vee L_n$ is true in $\mathbf{H}$ iff at least one L_i is true in $\mathbf{H}$;
- A clause C is true in $\mathbf{H}$ iff all ground instances of C arc truc in $\mathbf{H}$.

If every clause in a set of clauses $\mathbf{S}$ is true in $\mathbf{H}$, we say that $\mathbf{H}$ is a *Herbrand model* (or an *H-model*) of $\mathbf{S}$.

The correspondence between H-structure and F-structures can be stated as follows: Given an F-structure for a set of clauses $\mathbf{S}$, the corresponding H-structure is the set of round molecules that are true in the F-structure. Conversely, for an H-structure $\mathbf{H}$, the corresponding F-structure $\mathbf{I}_H = \langle U, \preceq_U, \in_U, \mathbf{I}_{\mathcal{F}}, \mathbf{I}_{\mathcal{P}}, \mathbf{I}_{\rightarrow}, \mathbf{I}_{\twoheadrightarrow}, \mathbf{I}_{\bullet\rightarrow}, \mathbf{I}_{\bullet\twoheadrightarrow}, \mathbf{I}_{\Rightarrow}, \mathbf{I}_{\Rightarrow\!\!\!\Rightarrow} \rangle$ is defined as follows:

1. The domain U is $U(\mathcal{F})/ \doteq$, i.e., the quotient of $U(\mathcal{F})$ induced by thc equalitics in $\mathbf{H}$. We denote the equivalence class of t by $[t]$.
2. The ordering $\preceq_U$ and the class membership relation $\in_U$ are determined by the is-a assertions in $\mathbf{H}$. For all $[t], [s] \in U$, we assert $[s] \preceq_U [t]$ iff $s{::}t \in \mathbf{H}$ and $[s] \in_U [t]$ iff $s : t \in H$.
3. $I_{\mathcal{F}}(c) = [c]$ for every 0-ary function symbol $c \in \mathcal{F}$.
4. $I_{\mathcal{F}}(f)([t_1], \ldots, [t_n]) = [f(t_1, \ldots, t_k)]]$ for every k-ary ($k \geq 1$) function symbol $f \in \mathcal{F}$.
5. $I^{(k)}_{\rightarrow}([scalM])([obj], [t_1], \ldots, [t_n])$
$$= \begin{cases} [s] & \text{if } obj[calM@t_1, \ldots, t_k \rightarrow s] \in \mathbf{H} \\ \text{undefined otherwise} \end{cases}$$
6. $I^{(k)}_{\rightarrow}([setM])([obj], [t_1], \ldots, [t_n])$
$$= \begin{cases} \{[s] \mid obj[setM@t_1, \ldots, t_k \twoheadrightarrow s] \in \mathbf{H}\} & \text{if } obj[setM@t_1, \ldots, t_k \twoheadrightarrow \{\ \}] \in \mathbf{H} \\ \text{undefined} & \text{otherwise} \end{cases}$$

The mappings $I_{\bullet\rightarrow}$ and $I_{\bullet\twoheadrightarrow}$ are defined similarly to (5) and (6), except that inheritable data expressions must be used instead of non-inheritable ones.

7. $I^{(k)}_{\Rightarrow}([obj], [t_1], \ldots, [t_n])$
$$= \begin{cases} \{[s] \mid obj[scalM@t_1, \ldots, t_k \Rightarrow s] \in \mathbf{H}\} & \text{if } obj[scalM@t_1, \ldots, t_k \Rightarrow (\,)\}] \in \mathbf{H} \\ \text{undefined} & \text{otherwise} \end{cases}$$
8. $I^{(k)}_{\Rrightarrow}([obj], [t_1], \ldots, [t_n])$
$$= \begin{cases} \{[s] \mid obj[scalM@t_1, \ldots, t_k \Rrightarrow s] \in \mathbf{H}\} & \text{if } obj[scalM@t_1, \ldots, t_k \Rrightarrow (\,)\}] \in \mathbf{H} \\ \text{undefined} & \text{otherwise} \end{cases}$$
9. $I_{\mathscr{P}}(p) = \{\langle [t_1], \ldots, [t_k] \rangle \mid p(t_1, \ldots, t_k) \in \mathbf{H}\}$

Proposition 6.2 *Let* **S** *be a set of clauses. Then,* **S** *is unsatisfiable iff* **S** *has no H-model.*

Proof It is easy to verify that for every H-structure **H**, the entailment $\mathbf{H} \models \mathbf{S}$ takes place iff $\mathbf{I}_H \models \mathbf{S}$, where $\mathbf{I}_H$ is the F-structure that corresponds to **H**.

In F-logic, the next theorem called the *Herbrand's theorem* holds. To present it, several concepts are needed. A set **S** of ground clauses is *finitely satisfiable* if every finite subset of **S** is satisfiable. A finitely satisfiable set **S** is *maximal* if no other set of ground clauses containing **S** is finitely satisfiable.

Lemma 6.10 *Given a finitely satisfiable set of ground clauses* **S***, there exists a maximal finitely satisfiable set* **T** *such that* $\mathbf{S} \subseteq \mathbf{T}$.

Proof Let Λ be a collection of all finitely satisfiable sets of ground clauses (in a fixed language $\mathscr{L}$) that contains **S**. The set Λ is partially ordered by set-inclusion. Since $\mathbf{S} \in \Lambda$, Λ is non-empty. Furthermore, for every $\subseteq$-growing chain $\Sigma \subseteq \Lambda$, the least upper bound of the chain $\bigcup \Sigma$ is also in Λ. Indeed:

- $\bigcup \Sigma$ contains **S**; and
- $\bigcup \Sigma$ is finitely satisfiable.

By Zorn's lemma, Λ has a maximal element.

Lemma 6.11 *Let* **T** *be a maximal finitely satisfiable set of ground clauses.*

(i) For every ground F-molecule T*, either* $T \in \mathbf{T}$ *or* $\neg T \in \mathbf{T}$.
(ii) A ground clause, $L_1 \vee \ldots \vee L_n$ *is in* **T** *iff* $L_o \in \mathbf{T}$ *for some* $1 \leq i \leq n$.

Proof (i): If **T** is finitely satisfiable, then so is $\mathbf{T} \cup \{T\}$ or $\mathbf{T} \cup \{\neg T\}$. Therefore, **T** must contain either T or $\neg T$, since it is maximal. (ii) is proved similarly to (i).

Lemma 6.12 *Let* **T** *be a maximal finitely satisfiable set of ground clauses. Let* **H** *be the set of all ground molecules in* **T**. *Then,* **H** *is an H-structure.*

Proof The set **H** is $\models$-closed, since so is **T** (or else **T** is not maximal).

Theorem 6.22 (Herbrand's theorem) *A set of clauses* **S** *is unsatisfiable iff so is some finite subset of ground instances of the clauses in* **S**.

Proof For the "if" part, assume that some finite subset of ground clauses of **S** is unsatisfiable. Then, **S** is also unsatisfiable.

The "only-if" part is proved by contradiction. Let $\mathbf{S}'$ denote the set of all ground instances of the clauses in **S**, and suppose that all finite subsets of $\mathbf{S}'$ are satisfiable, which implies that $\mathbf{S}'$ itself is finitely satisfiable. We will show that **S** is satisfiable.

By Lemma 6.10, $\mathbf{S}'$ can be extended to a maximal finitely satisfiable set **T**. Let **H** be the set of all ground molecules in **T**, which is an H-structure, by Lemma 6.12. We claim that $\mathbf{H} \models C$ iff $C \in \mathbf{T}$ for every ground clause C. Consider the following cases.

(a) C s a ground molecule. By definition, $\mathbf{H} \models C$ iff $C \in \mathbf{T}$.
(b) C is a negative literal $\neg P$. Then, $\mathbf{H} \models \neg P$ iff $P \notin \mathbf{H}$. Since **H** contains all the ground molecules in **T**, $P \notin \mathbf{H}$ iff $P \notin \mathbf{T}$. Finally, by (i) of Lemma 6.11, $P \notin \mathbf{T}$ iff $\neg P \in \mathbf{T}$.
(c) C is a disjunction of ground literals $L_1 \vee \ldots \vee L_n$. Then, $\mathbf{H} \models L_1 \vee \ldots \vee L_n$ iff $\mathbf{H} \models L_i$ for some i, by definition
$L_i \in \mathbf{T}$ by case (a) and (b) above
iff $L_1 \vee \ldots \vee L_n \in \mathbf{T}$ by (ii) of Lemma 6.11.

We have shown that **H** satisfies every clause of **T**. Since $\mathbf{S}' \subseteq \mathbf{T}$, **H** is an H-model of **S**. By Proposition 6.2, **S** is satisfiable.

Next, we consider substitutions in F-logic. Let $\mathscr{L}$ be a language with a set of variables $\mathscr{V}$. A *substitution* is a mapping $\sigma \to \{\text{id} - \text{terms of } \mathscr{L}\}$ such that it is an identity everywhere outside some finite set $dom(\sigma) \subseteq \mathscr{V}$, the *domain* of σ. As in classical logic, substitutions extend to mappings $\{id - terms\} \to \{id - terms\}$ as follows:

$$\sigma(f(t_1, \ldots, t_n)) = f(\sigma(t_1), \ldots, \sigma(t_n))$$

A substitution σ can be further extended to a mapping from molecules to molecules by distributing σ through molecules' components.

A substitution is *ground* if $\sigma(X) \in U(\mathscr{F})$ for each $X \in dom(\sigma)$, that is, if $\sigma(X)$ has no variables. Given a substitution σ and a formula φ, $\sigma(\varphi)$ is called an *instance* of φ. It is a *ground instance* if it contains no variables. A formula is *ground* if it has no variables.

Unification of id-terms, is-a molecules, and P-molecules is no different from that in classical logic. Let T_1 and T_2 be a pair of id-terms, is-a molecules, or P-molecules. A substitution σ is a *unifier* of T_1 and T_2 if $\sigma(T_1) = \sigma(T_2)$. This unifier is called a *most general unifier*, written $mgu(T_1, T_2)$, if for every unifier μ of T_1 and T_2, there exists a substitution γ such that $\mu = \gamma \circ \sigma$.

For object molecules, instead of requiring identity under unification we merely ask that a unifier would map molecules into *submolecules* of other molecules.

Definition 6.47 (*Asymmetric unification of object molecule*) Let $L_1 = S[\ldots]$ and $L_2 = S[\ldots]$ be a pair of object molecules with the same object id, S.

We say that L_1 is a *submolecule* of L_2, denotes $L_1 \sqsubseteq L_2$ iff every constituent atom of L_1 is also a constituent atom of L_2.

A substitution σ is a *unifier* of L_1 into L_2 iff $\sigma(L_1) \sqsubseteq \sigma(L_2)$.

To define most general unifiers for object molecules, we must consider complete sets of most general unifiers. Consider $L_1 = a[set \twoheadrightarrow X]$ and $L_2 = a[set \rightarrow \rightarrow \{b, c\}]$. Intuitively, there are two unifiers of L_1 into L_2 that can be called "most general": b/X and c/X. Clearly, none of these unifiers is more general than the other and, therefore, the definition of mgu that works for P-molecules and for is-a assertions does not work.

Definition 6.48 (*Most general unifiers*) Let L_1 and L_2 be a pair of molecules and let α, β be a pair of unifiers of L_1 into L_2. We say that α is *more general* than β, denoted $\alpha \trianglelefteq \beta$, iff there is a substitution γ such that $\beta = \gamma \circ \alpha$. A unifier α of L_1 into L_2 is a *most general unifier* (mgu) if for every unifier β, $\beta \trianglelefteq \alpha$ implies $\alpha \trianglelefteq \beta$.

A set Σ of most general unifiers of L_1 into L_2 is *complete* if for every unifier θ of L_1 into L_2 there is $\alpha \in \Sigma$ such that $\alpha \trianglelefteq \theta$.

The complete set of unifiers of L_1 into L_2 is unique up to the equivalence, as in the classical case. In predicate calculus, the notion of a unifier works for an arbitrary number of terms to be unified. Extension of the above definition to accommodate an arbitrary number of id-terms, P-molecules or is-a assertions is obvious. For object molecules, we say that a substitution σ is a *unifier* of $L_1, \ldots, L_n$ into L when $\sigma(L_i) \sqsubseteq L$ for $i = 1, \ldots, n$. Generalization of the notion of mgu is straightforward.

For convenience, we also define mgu's for tuples of id-terms. Tuples $\langle P_1, \ldots, P_n \rangle$ and $\langle Q_1, \ldots, Q_n \rangle$ are *unifiable* when there is a substitution σ such that $\sigma(P_i) = \sigma(Q_i), i = 1, \ldots, n$. This unifier is *most general*, written $mgu(\langle P_1, \ldots, P_n \rangle, \langle Q_1, \ldots, Q_n \rangle))$, if for every other unifier μ of these tuples $\mu = \gamma \circ \sigma$ for some substitution γ. It is easy to see that any mgu of $\langle P_1, \ldots, P_n \rangle$ and $\langle Q_1, \ldots, Q_n \rangle$ coincides with the mgu of $f(P_1, \ldots, P_n)$ and $f(Q_1, \ldots, Q_n)$, where f is some n-ary function symbol. Therefore, mgu of a pair of tuples is unique.

We are now ready to describe core inference rules. For simplicity, only binary resolution is considered. In these inference rules, the symbols L and L' denote positive literals, C and C' demote clause, and P, Q, R, S etc. denote id-terms.

Resolution: Let $W = \neg L \vee C$ and $W' = L' \vee C'$ be a pair of clauses that are standardized apart. Let θ be an mgu of L into L'. The *resolution* rule is as follows:

from W and W' derive $(C \vee C')\theta$

Notice that when L and L' are object molecules, resolution is *asymmetric* since $\theta = mgu_{\sqsubseteq}(L, L')$ may be different from $mgu_{\sqsubseteq}(L', L)$, and the latter mgu may not even exist.

As in the classical case, binary resolution must be complimented with the so-called factoring rule that seeks to reduce the number of disjuncts is in a clause.

Factoring: The *factoring* rule has two forms, depending on the polarity of literals to be factored. For positive literals, consider a clause of the form $W = L \vee L' \vee C$, where L and L' are positive literals. Let L be unifiable into L' with the mgu θ. The factoring rule is as follows:

from W derive $(L \vee C)\theta$

In the case of negative literals, $W = \neg L \vee \neg L' \vee C$ and L is unifiable into L' with the mgu θ, then the factoring rule is:

from W derive $(\neg L' \vee C)\theta$

Clauses inferred by one of the two factoring rules are called *factors* of W. Note that in both inference rules L must be unifiable into L'. However, in the first case, it is the literal L that survives, while in the second rule it is L'.

To account for the equality relation, we need a paramodulation rule. When there is a need to focus on a specific occurrence of an id-term T in an expression E, it is a standard practice to write E as $E[T]$. If one single occurrence of T is replaced by S, the result will be denoted by $E[S/T]$.

Paramodulation: Consider a pair of clauses, $W = L[T] \vee C$ and $W' = (T' \doteq T'') \vee C'$, with no common variables. If T and T' are id-terms unifiable with an mgu θ, then the *paramodulation* rule says:

from W and W' derive $(L[T''/T] \vee C \vee C')\theta$

Next, we consider is-a inference rules. The following axiom and rules capture the semantics of the subclass-relationship and its interaction with class membership.

IS-A reflexivity: The following is the *IS-A reflexivity* axiom:

$(\forall X)X{::}X$

IS-A acyclicity: Let $W = (P{::}Q) \vee C$ and $W' = (Q'{::}P') \vee C'$ be clauses with no variables in common. Suppose that θ is an mgu o tuples $\langle P, Q\rangle$ and $\langle P', Q'\rangle$ of id-terms. The *IS-A acyclicity* rule is as follows:

from W and W' derive $((P \doteq Q) \vee C \vee C')\theta$

Note that IS-A reflexivity and IS-A acyclicity imply reflexivity of equality. Indeed, since $X{::}X$ is an axiom, by IS-A acyclicity, one can derive $X \doteq X$ from $X{::}X$ and $X{::}X$:.

IS-A transitivity: Let $(P{::}Q) \vee C$ and $W' = (Q'{::}R') \vee C'$ be standardized apart, and let θ be an mgu of Q and Q'. The *IS-A transitivity* rule is:

from W and W' derive $((P{::}R') \vee C \vee C')\theta$

Subclass inclusion: Let $(P{::}Q) \vee C$ and $W' = (Q'{::}R') \vee C'$ be standardized apart, and let θ be an mgu of Q and Q'. The *subclass inclusion* rule is:

from W and W' derive $((P : R') \vee C \vee C')\theta$

Signature expressions have the properties of type inheritance, input, restriction, and output relaxation that are captured by the following inference rules:
Type inheritance: Let $W = P[Mthd@Q_1, \ldots, Q_k \Rightarrow T] \vee C$ and $W' = (S'::P') \vee C'$ be a pair of clauses with common variables, and suppose P and P' have an mgu θ. The *type inheritance* rule states the following:

from W and W' derive $(S'[Mthod@Q_1, \ldots, Q_k \Rightarrow T] \vee C \vee C')\theta$

In other words, S' inherits the signature of P'. A similar rule exists for set-valued methods. If $W = P[Mthod@Q_1, \ldots, Q_k \Rrightarrow T] \vee C$ and W' is as before, then:

from W and W' derive $(S'[Mthod@Q_1, \ldots, Q_k \Rrightarrow T] \vee C \vee C')\theta$

Input restriction: Let $W = P[Mthd@Q_1, \ldots, Q_k \Rightarrow T] \vee C$ and $W' = (Q_i''''::Q_i') \vee C'$ be standardized apart. Suppose also that Q_i and Q_i' have an mgu θ. The *input restriction* rule states:

from W and W' derive $(P[Mthod@Q_1, \ldots, Q_i'', \ldots Q_k \Rightarrow T] \vee C \vee C')\theta$

Here, Q_i'' replaces Q_i. A similar rule exists for set-valued methods. It should also be noted that T in the above three inference rules stands for "()" or an id-term.
Output relaxation: Consider clauses $W = P[Mthd@Q_1, \ldots, Q_k \Rightarrow R] \vee C$ and $W' = (R'::R'') \vee C'$ with no common variables, suppose R and R' have an mgu θ. The *output relaxation* rule sates:

from W and W' derive $(P[Mthd@Q_1, \ldots, Q_k \Rightarrow R''] \vee C \vee C')\theta$

A similar rule applies to set-valued methods.

There are miscellaneous inference rules. The requirement that a scalar method must return at most one value is built into the following rule:
Scalarity: Consider a pair of clauses that share no common variables:

$$W = P[Mthd@Q_1, \ldots, Q_k \rightarrow R] \vee C$$
$$W' = P'[Mthd'@Q'_1, \ldots, Q'_k \rightarrow R'] \vee C'.$$

Suppose there is an mgu θ that unifies the tuple of id-terms $\langle P, Mthd, Q_1, \ldots, Q_k\rangle$ with the tuple $\langle P', Mthd', Q'_1, \ldots, Q'_k\rangle$. The rule of *scalarity* says:

from W and W' derive $((R \doteq R') \vee C \vee C')\theta$

A similar rule exists for inheritable scalar expressions. The only difference is that $\rightarrow$ is replaced with $\bullet\!\!\rightarrow$ in W and W'.

Another miscellaneous rule is called *merging*. It seeks to combine information contained in different molecules. Let L_1 and L_2 be a pair of such molecules with the same object id. An object-molecule L is called a merge of L_1 and L_2, if the set of constituent atoms of L is precisely the union of the sets of constituent atoms of L_1 and L_2. A pair of molecules can be merged in several different ways when they have common set-valued methods.

A *canonical merge* of L_1 and L_2, denoted $merge(L_1, L_2)$, is a merge that does not contain repeated identical invocations of set-valued methods.

Merging: Consider a pair of standardized apart clauses $W = L \vee C$ and $W' = L' \vee C'$, where L and L' are object molecules. Let θ be an mgu unifying the object id parts of L and L'. Let L'' denote the canonical merge of $L\theta$ and $L'\theta$. The merging rule says:

from W and W' derive $L'' \vee (C \vee C')\theta$

Finally, since for every id-term P, the molecule $P[]$ is a tautology, we have the following *elimination* rule.
Elimination: If C is a clause and P an id-term then:

from $\neg P[] \vee C$ derive C

Notice that if C is an empty clause then the elimination rule would derive an empty clause as well.

Now, we show the soundness of the proof theory. Given a set **S** of clauses, a *deduction* of a clause C from **S** is a finite sequence of clauses $D1, \ldots, D_n$ such that $D_n = C$ where, for $1 \leq k \leq n$, D_k is either

- a member of **S**, or
- is derived from some D_i, and, possibly, an additional clause D_j, where $i, j < k$, using one of the core, is-a, type, or miscellaneous inference rules.

A deduction ending with the empty clause $\square$, is called a *refutation* of **S**. If C is deducible from **S**, we write $\mathbf{S} \vdash C$.

Theorem 6.23 (Soundness of F-logic deduction) *If* $\mathbf{S} \vdash C$ *then* $\mathbf{S} \models C$.

Proof Directly follows from the closure properties given above and from t form of the inference rules.

We here give a sample proof.

Example 6.6 Consider the following set of clauses:

(i) $a::b$
(ii) $p(a)$
(iii) $c[m@b \Rightarrow (v, w)]$
(iv) $r[attr \rightarrow a]$
(v) $r[attr \rightarrow f(S)] \vee \neg p(X) \vee \neg O[M@X \Rightarrow S]$
(vi) $\neg p(f(Z))$

We can refute the above set using the following sequence of deduction steps, where θ denotes the unifier used in the corresponding step:

(vii) $r[attr \rightarrow f(S)] \vee \neg O[M@q \Rightarrow S]$ (by resolving (ii) and (v); $\theta = \{a/X\}$)
(viii) $c[m@a \Rightarrow (v, w)]$ (by input restriction from (i) and (iii))
(ix) $r[attr \rightarrow f(v)]$ (by resolving (vii) with (vii); $\theta = \{c/O, v/S, m/M\}$)

(x) $a \doteq f(v)$ (by the rule of scalarity, using (iv) and (ix))

(xi) $p(f(v))$; (by paramodulation, using (ii) and (x))

(xii) $\square$ (by resolving (vi) with (xi); $\theta = \{v/Z\}$)

Next, we state the completeness theorem of F-logic. First, we establish completeness for the ground case by Herbrand's theorem. Then, by an analogue of the Lifting Lemma, the completeness of ground refutations can be generalized for the non-ground case.

Lemma 6.13 *Let* **S** *be a set of ground F-literals. If* **S** *is unsatisfiable then there are molecules* P *and* Q *such that* $P \sqsubseteq Q, \neg P \in \mathbf{S}$ *and* $\mathbf{S} \vdash Q$. *(When,* P, Q *are P-molecules or is-a assertions,* $P \sqsubseteq Q$ *should be taken to mean that* P *is identical to* Q.

Proof Suppose, to the contrary, that there are no molecules P and Q such that $P \sqsubseteq Q, \neg P \in \mathbf{S}$, and $\mathbf{S} \vdash Q$. We will show that **S** must be satisfiable. Consider the following set of molecules:

$\mathbf{D}(\mathbf{S}) =_{def} \{P \mid P$ is a submolecule of some molecule Q such that $\mathbf{S} \vdash Q\}$.

Since every H-structure **H** has a corresponding F-structure $\mathbf{I}_H$ such that $\mathbf{H} \models \mathbf{S}$ iff $\mathbf{I}_H \models \mathbf{S}$, from **D**(**S**), as explained above, we can obtain an F-structure **M**. Since **D**(**S**) is closed under deduction, **M** is shown to be an F-structure.

We claim that for every molecule P:

$\mathbf{M} \models P$ iff $P \in \mathbf{D}(\mathbf{S})$

The "if"-direction follows from soundness of the derivation rules. For the "only if"-direction, assume $\mathbf{M} \models P$ and consider the following cases.

(i) P is an is-a assertion or a predicate:
If P is an is-a assertion, by Definition 6.46(2) in the construction of **M** can be used to show that $\mathbf{M} \models P$ iff $P \in \mathbf{D}(\mathbf{S})$. If P is a predicate, Definition 6.46(9) can be used to show the same.

(ii) P is an abject molecule composed of atoms $\tau_1, \ldots, \tau_n$.
Then, $\mathbf{M} \models P$ iff $\mathbf{M} \models \tau_i, i = 1, \ldots, n$. By Definition 6.46(5), (6), (7) or (8), it follows that $\mathbf{M} \models \tau_i$ iff $\tau_i \in \mathbf{D}(\mathbf{S})$. Therefore, by the definition of **D**(**S**), there are molecules $Q_1, \ldots, Q_n$ deducible from **S** such that τ_i i a submolecule of Q_i for $i = 1, \ldots, n$. Let Q be the canonical merge of $Q_1, \ldots, Q_n$.
Then, P is a submolecule of Q since every constituent atom of P is also a constituent atom of Q, and Q is deducible from **S** since $Q_1, \ldots, Q_n$ are deducible from **S** and Q is a merge of $Q_1, \ldots, Q_n$. Hence, P is in **D**(**S**).

By the definition of **D**(**S**), if $P \in \mathbf{S}$ is a positive literal then $P \in \mathbf{D}(\mathbf{S})$. Hence, by Definition 6.46(7), $\mathbf{M} \models P$. For every negative literal $\neg P$ in **S**, P is not a molecule of any molecule deducible from **S**, by the assumption of the present proof. So, P is not in **D**(**S**), by Definition 6.46(7), $\mathbf{M} \not\models P$ and therefore $\mathbf{M} \models \neg P$. Thus, **M** satisfies every literal of **S**, that is, it is a model for **S**.

Theorem 6.24 (Completeness of ground deduction) *If a set of ground clauses* **S** *is unsatisfiable, then there exists a refutation of* **S**.

Proof By Herbrand's theorem, we can assume that **S** is finite. Suppose **S** is unsatisfiable. We will show that there is a refutation of **S**. The proof is carried out by induction on the parameter *excess*(**S**), i.e., the number of excess literals in **S**:

$excess(\mathbf{S}) =_{def}$ (the number of occurrences of literals in **S**) – (the number of clauses in **S**).

Basis: $excess(\mathbf{S}) = 0$. The number of clauses in **S** equals the number of occurrences of literals in **S**. Hence, either $\Box \in \mathbf{S}$ and we are done, or every clause in **S** is a literal. In the latter case, by Lemma 6.13, $\mathbf{S} \vdash \neg P$ and $\mathbf{S} \vdash Q$ for some molecule P, Q such that $P \sqsubseteq Q$. Applying the resolution rule to $\neg P$ and Q, we obtain the empty clause.

Inductive step: $excess(\mathbf{S}) = n > 0$. There must be a clause C in **S** that contains more than one literal. Let us distinguish this clause from other clauses and write $\mathbf{S} = \{C\} \cup \mathbf{S}'$, where $C = L \vee C'$. Here, $C' \neq \Box$, since we have assumed that C contains more than one literal. By the distributivity law, $\{L \vee C'\} \cup \mathbf{S}'$ is unsatisfiable iff so are $\mathbf{T}_1 = \{C'\} \cup \mathbf{S}'$ and $T_2 = \{L\} \cup \mathbf{S}'$. Since $excess(\mathbf{T}_1) < n$ and $excess(\mathbf{T}_2) < n$, the induction hypothesis ensures that there are refutations of $\mathbf{T}_1$ and $\mathbf{T}_2$ separately. Therefore, $\mathbf{T}_1 \vdash \Box$. Let $dedseq_1$ denote the deduction that derives $\Box$ from $\mathbf{T}_1$.

Applying the deductive steps in $dedseq_1$ to **S**, we would derive either L or $\Box$. If $\Box$ is produced, then **S** is refuted. Otherwise, if L is produced, it means that $\mathbf{S} \vdash L$. Let $dedseq_2$ denote the derivation that refutes $\mathbf{T}_2 = \{L\} \cup \mathbf{S}'$ (which exists by the inductive assumption). Since $\mathbf{S}' \subseteq \mathbf{S}$ and $\mathbf{S} \vdash L$, it follows that if we apply $deedseq_1$ to **S** and then follow this up with steps from $dedseq_2$, we refute **S**.

Proposition 6.3 *There exists a unification algorithm that, given a pair of molecules* T_1 *and* T_2, *yields complete set of mgu's of* T_1 *and* T_2.

Lemma 6.14 (Lifting lemma) *Suppose* C_1, C_2 *are clauses and* ${C_1}', {C_2}'$ *are their instances, respectively. If* D' *is derived from* ${C_1}'$ *and* ${C_2}'$ *(or from* ${C_1}'$ *alone) using the derivation rules, then there exists a clause D such that*

- *D is derivable from the factors of* C_1 *and* C_2 *(or from* C_1 *alone) via a single derivation step;*
- *This derivation step uses the same inference rule as the one that derived* D'*; and*
- D' *is an instance of D.*

Proof Consider each derivation rule separately. The proof in each case is similar to the corresponding proof in predicate calculus, since the notion of substitution is the same in both logics.

Theorem 6.25 (Completeness of F-logic inference system) *If a set* **S** *of clauses is unsatisfiable, then there is a refutation of* **S**.

Proof Consider $\mathbf{S}^*$, the set of all ground instances of $\mathbf{S}$. By the ground case (Theorem 6.24), there is a refutation of $\mathbf{S}^*$. By the Lifting lemma, this refutation can be lifted to refutation of $\mathbf{S}$.

Kifer et al. [97] further dealt with data modeling, well-typed programs and type errors, encapsulation, and inheritance in F-logic; also see Kifer and Wu [102].

Although a starting point of F-logic is similar to that of annotated logics, their exposition of F-logic does not address the representation of inconsistency. They recognized the point, and suggested a possibility of combining F-logic and annotated logics.

6.4 Neural Computing

Neural Computing is a branch which studies computational models inspired by biological neural networks. The theoretical basis for neural computing is the so-called *Artificial Neural Network* (ANN), and has been extensively studied in AI.

An ANN can be described as a computational system consisting of a set of highly interconnected processing elements, called *artificial neurons*, which process information as a response to external stimuli. An artificial neuron is a simplistic representation that emulates the signal integration and threshold firing behavior of biological neurons by means of mathematical structures.

Like their biological counterpart, artificial neurons are bound together by connections that determine the flow of information between peer neurons. Stimuli are transmitted from one processing element to another via synapses or interconnections, which can be excitatory or inhibitory; see Fausett [81] for details.

The advantage of neural networks over conventional programming lies on their ability to solve problems that do not have an algorithmic solution or the available solution is too complex to be found.

Thus, neural networks are well suited to tackle problems that people are good at solving, like prediction and pattern recognition. In fact, neural networks have been applied to the fields such as medical domain, engineering, automation, robotics, etc.

Several theories of artificial neural networks have been proposed with different characteristics. Annotated logics $E\tau$ can provide a new theory of ANN, i.e., *paraconsistent Artificial Neural Networks* (PANN), introduced by Da Silva Filho and Abe [5, 70].

$E\tau$ are a class of annotated logics for evidential reasoning. They are interpreted as a variant of $P\tau$. Atomic formulas of $E\tau$ are of the type $p_{(\mu,\lambda)}$, where $(\mu, \lambda) \in [0, 1]^2$ which is a lattice with the ordering $(\mu_1, \lambda_1) \leq (\mu_2, \lambda_2) \Leftrightarrow \mu_1 \leq \mu_2$ and $\lambda_1 \leq \lambda_2$.

Table 6.1 Extreme and non-extreme states

Extreme states	Symbol	Non-extreme states	Symbol
True	V	Quasi-true tending to inconsistent	$QV \rightarrow T$
False	F	Quasi-true tending to paracomplete	$QV \rightarrow \perp$
Inconsistent	T	Quasi-false tending to inconsistent	$QF \rightarrow T$
Paracomplete	$\perp$	Quasi-false tending to paracomplete	$QF \rightarrow \perp$
		Quasi-inconsistent tending to true	$QT \rightarrow V$
		Quasi-inconsistent tending to false	$QT \rightarrow F$
		Quasi-paracomplete tending to true	$Q\perp \rightarrow V$
		Quasi-paracomplete ending to false	$Q\perp \rightarrow F$

Based on $E\tau$, we introduce the uncertainty degree G_{un} and the certainty degree G_{ce} as follows:

$$G_{un}(\mu, \lambda) = \mu + \lambda - 1$$
$$G_{ce}(\mu, \lambda) = \mu - \lambda$$

We consider the following 12 output states in Table 6.1:
The control values are:

- V_{cic} = maximum value of uncertainty control = Ft_{ct}
- V_{cve} = maximum value of certainty control = Ft_{ce}
- V_{cpa} = minimum value of uncertainty control = $-Ft_{ct}$
- V_{cpa} = minimum value of certainty control = $-Ft_{ce}$

For the present discussion, we will use the following: $Ft_{ct} = Ft_{ce} = \frac{1}{2}$.
We also set $C_1 = C_3 = \frac{1}{2}$ and $C_2 = C_4 = -\frac{1}{2}$.
All states are represented in Fig. 6.1.
In the paraconsistent annotated structure, the main aim is to know how to measure or to determine the certainty degree concerning a proposition, if it is False or True. Therefore, we take into account only the certainty degree G_{ce}.

The uncertainty degree G_{un} indicates the measure of the inconsistency or paracompleteness. If the certainty degree is low or the uncertainty degree is high, it generates an indefinition.

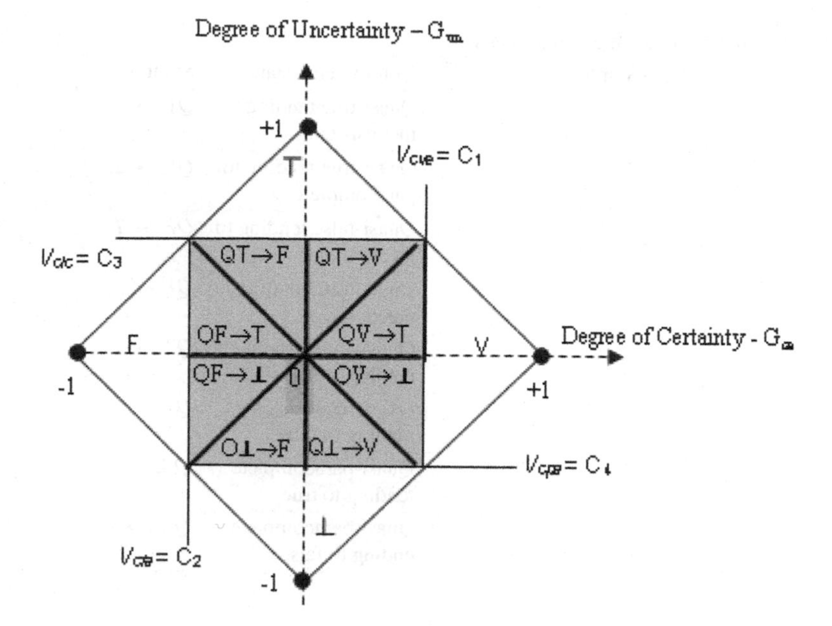

Fig. 6.1 Representation of the certainty and uncertainty degrees

Definition 6.49 (*Certainty degree*) The resulting certainty degree G_{ce} is obtained as follows:

if: $V_{cfa} \leq G_{un} \leq G_{cve}$ or $V_{cic} \leq G_{un} \leq G_{cpa} \Rightarrow G_{ce} =$ Indefinition
For: $V_{cpa} \leq G_{un} \leq V_{cic}$
if: $G_{un} \leq V_{cfa} \Rightarrow G_{ce} =$ False with degree G_{un}
The algorithm that expresses a basic Paraconsistent Artificial Neural Cell - PANC - is:
*/ Definition of the adjustable values */
$V_{cve} = C_1$ /* maximum value of certainty control */
$V_{cfa} = C_2$ /* minimum value of certainty control */
$V_{cic} = C_3$ /* maximum value of uncertainty control */
$V_{cpa} = C_4$ /* minimum value of uncertainty control */
/* Input Variables */
μ, λ
/* Output Variables */
Digital output $= S1$
Analog output $= S2a$
Analog output $= S2b$
/* Mathematical expressions */
begin:
$0 \leq \mu \leq 1$ and $0 \leq \lambda \leq 1$
$G_{un} = \mu + \lambda - 1$

$G_{ce} = \mu - \lambda$
/* determination of the extreme states */ if $G_{ce} \geq C_1$ then $S_1 = V$
if $G_{ce} \geq C_2$ then $S_1 = F$
if $G_{un} \geq C_3$ then $S_1 = T$
if $G_{un} \leq C_4$ then $S_1 = \perp$
if not: $S_1 = I$ – Indetermination
$G_{un} = S2a$
$G_{ce} = S2b$

A PANC is called *basic PANC* when given a pair (μ, λ) is used as an input and resulting as the output G_{un} = resulting uncertainty degree, G_{ce} = resulting certainty degree, and X = constant of Indefinition, calculated by the equations $G_{un} = \mu + \lambda - 1$ and $G_{ce} = \mu - \lambda$.

The basic paraconsistent artificial neural cell has the structure as Fig. 6.2.

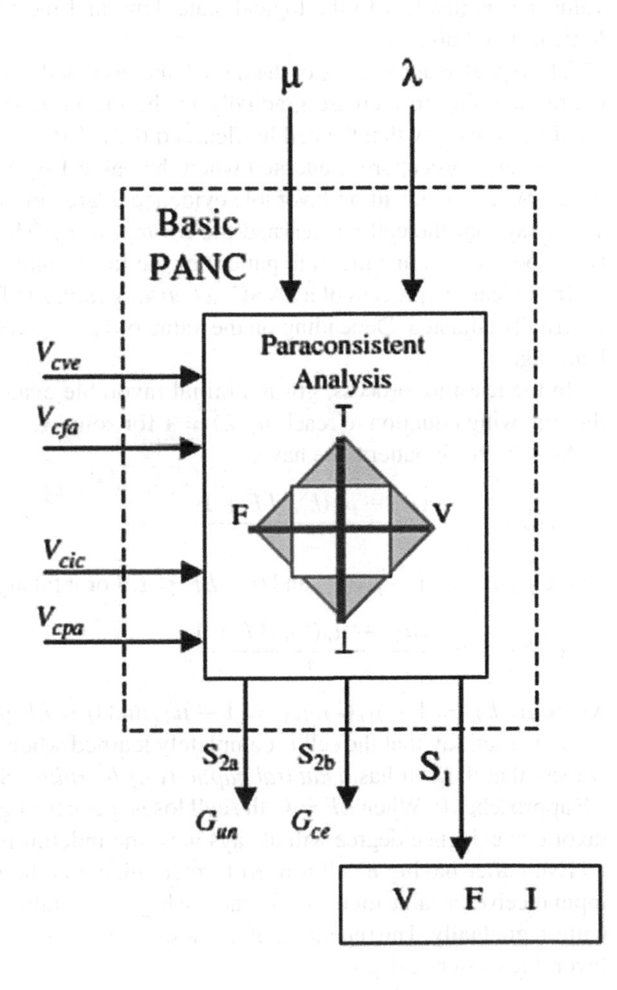

Fig. 6.2 The basic paraconsistent artificial neural cell

Now, we present a *paraconsistent Artificial Neural Cell of Learning* (PANC-l), which is obtained from a PANC. In this learning cell, sometimes we need the action of the operator Not in the training process. Its function is to perform the logical negation in the resulting output sign. For a training process, we consider initially a PANC of Analytic Connection which does not undergoing any learning process.

According to the paraconsistent analysis, a cell in these conditions has two inputs with an indefinite value $\frac{1}{2}$. So, the basic structural equation yields the same value $\frac{1}{2}$ as output, having as result an indefinition.

The learning cells can be used in the PANC as memory units and pattern sensors in preliminary layers. For instance, a PANC-l can be trained to learn a pattern by using an algorithm. For the training of a cell we can use as a pattern real values between 0 and 1. The cells can also be trained to recognize values between 0 and 1.

The learning of the cells with extreme values 0 or 1 composes the primary sensorial cells. Thus, the primary sensorial cells consider as pattern a binary digit, where the value 1 is equivalent to the logical state True and the value 0 is equivalent to the logical state False.

The repeated appearance of the input 0 means that the resulting favorable evidence degree is going to increase gradually in the output reaching the value 1. In these conditions, we say that the cell has learned the *falsity pattern*.

The same procedure is adopted when the value 1 is applied to the input repeated times. When the resulting favorable evidence degree in the output reaches the value 1, we say that the cell has learned the *truth pattern*. Therefore, a PANC can learn two types of patterns: the truth pattern or the falsity pattern.

In the leaning process of a PANC, a *learning factor* (LF) can be introduced that is externally adjusted. Depending on the value of LF, it gives the cell a faster or slower learning.

In the learning process, given a initial favorable evidence degree $\mu_r(k)$, we use the following equation to reach $\mu_r(k) = 1$ for some k.

So for a truth pattern, we have:

$$\mu_r(k+1) = \frac{(\mu_1 - \mu_r(k)_c)LF + 1}{2}$$

where $\mu_r(k)_c = 1 - \mu_r(k)$, and $0 \leq LF \leq 1$. For a falsity pattern, we have:

$$\mu_r(k+1) = \frac{(\mu_{1c} - \mu_r(k)_c)LF + 1}{2}$$

where $\mu_r(k)_c = 1 - \mu_r(k)$, $\mu_{1c} = 1 - \mu_1$, and $0 \leq LF \leq 1$.

So we can say that the cell is completely learned when $\mu_r(k+1) = 1$. If $LF = 1$, we say that the cell has a *natural capacity of learning*. Such capacity decreases as LF approaches 0. When $LF = 0$, the cell loses the learning capacity and the resulting favorable evidence degree will always have the indefinition value $\frac{1}{2}$.

Even after having a cell trained to recognize a certain pattern, if insistently the input receives a value totally different, the high uncertainty makes the cell unlearn the pattern gradually. The repetition of the new values implies a decrease of the resulting favorable evidence degree.

Then, the analysis has reached an indefinition. By repeating this value, the resulting favorable evidence degree reaches 0, meaning that the cell is giving the null favorable evidence degree to the former proposition to be learned.

This is equivalent to saying that the cell is giving the maximum value to the negation of the proposition, so the new pattern must be confirmed.

Algorithmically, this is showed when the certainty degree G_{ce} reaches the value -1. In this condition the negation of the proposition is confirmed. This is obtained by applying the operator NOT to the cell.

It inverts the resulting favorable evidence degree in the output. From this moment on the PANC considers as a new pattern the new value that appeared repeatedly and unlearning the pattern previously.

By considering two factors, LF—learning factor and UF—unlearning factor, the cell can learn or unlearn faster or slower according to the application. These factors are important in giving the PANN a more dynamic process.

Using the concepts of the basic PANC, we can obtain the family of PANC considered here, as described in Table 6.2.

Now, we show some basic aspects of how the PANN operates. Let us take three vectors:

$V_1 = (8, 5, 4, 6, 1),$
$V_2 = (8, 6, 4, 6, 5),$
$V_3 = (8, 2, 4, 6, 9)$

The favorable evidence is calculated as follows: given a pair of vectors, we take '1' for equal elements and '0' for the different elements, and we figure out its percentage.

Comparing V_2 with $V_1 : 1 + 0 + 1 + 1 + 0 = 3;$
as a percentage: $(3/5) * 100 = 60\,\%$

Comparing V_3 with $V_1 : 1 + 0 + 1 + 1 + 0 = 3;$
as a percentage: $(3/5) * 100 = 60\,\%$

Table 6.2 Paraconsistent artificial neural cells

PANC	Inputs	Calculations	Output
Analytic connection	μ, λ	$\lambda_c = 1 - \lambda$, G_{un}, G_{ce}	If $\mid G_{ce} \mid > Ft_{ce}$ then $S_1 = \mu_r$ and $S_2 = 0$
PANNac	Ft_{ct}, Ft_{ce}	$\mu_r = (G_{ce} + 1)/2$	If $\mid G_{ce} \mid > Ft_{ct}$ and $\mid G_{uu} \mid > \mid G_{ce} \mid$ then $S_1 = \mu$ and $S_2 = \mid G_{un} \mid$, if not $S_1 = 1/2$ and $S_2 = 0$
Maximization	μ, λ	None	If $\mu > \lambda$ then $S_1 = \mu$, if not $S_1 = \lambda$
PANNmax			
Minimization	μ, λ	None	If $\mu < \lambda$ then $S_1 = \mu$, if not $S_1 = \lambda$
PANNmin			

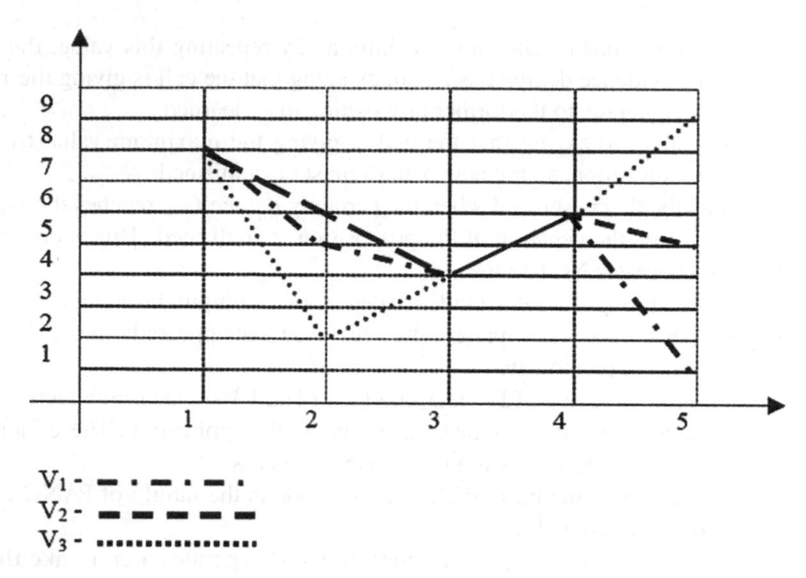

Fig. 6.3 Vectors V_1, V_2, V_3

The contrary evidence is the weighted addition of the differences between the different elements in the module:

$$\text{Comparing } V_2 \text{ with } V_1 : 0 + 1/10 + 0 + 0 + 4/10 = (5/10)/5 = 10\,\%$$
$$\text{Comparing } V_3 \text{ with } V_1 : 0 + 3/10 + 0 + 0 + 8/10 = (11/10)/5 = 22\,\%$$

Therefore, visually we see that vector V_2 is 'more similar' to V_1 than V_3. We use a PANN to recognize this similarity mathematically.

Vectors V_1, V_2 and V_3 are depicted as in Fig. 6.3.

Representing the learning of the vectors:

$$V_1 = (8, 5, 4, 6, 1)$$
$$V_2 = (8, 6, 4, 6, 5)$$
$$V_3 = (8, 2, 4, 6, 9)$$

in layers as in Fig. 6.4.

The table of the analyzed waves is as Table 6.3.

Comparing the analyzed wave and learned wave1, we have Table 6.4.

Comparing the analyzed wave and learned wave2, we have Table 6.5.

Normalizing the values by the division of the favorable evidence (μ) and of the contrary evidence (λ) for the number of elements of the wave, we have Table 6.6.

Therefore, we notice that the wave with the maximum favorable evidence and the minimum contrary evidence is the learned wave1, in other words, this is the most similar wave to the analyzed wave.

A problem in *EEG* (Electroencephalography) analysis, as well as any other *measurement* device is the inherent imprecision of the several sources involved:

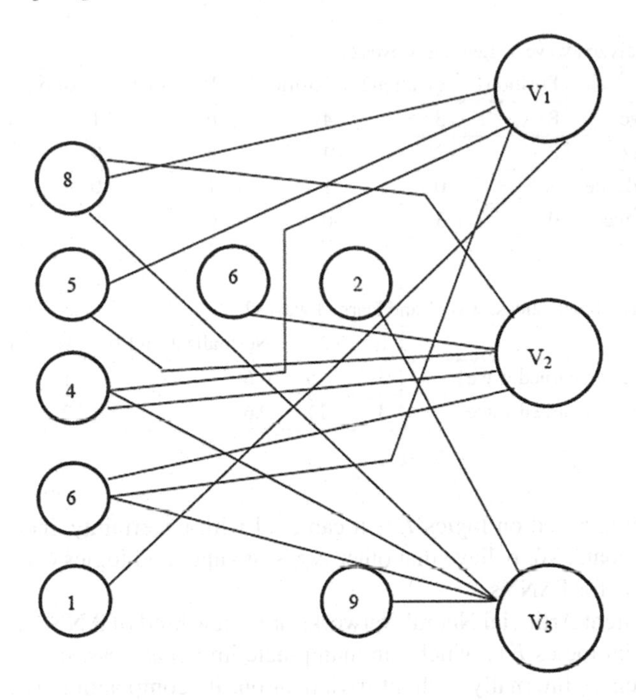

Fig. 6.4 PANN and layers

Table 6.3 Table of the analyzed waves

Wave's name	Position1	Position2	Position3	Position4	Position5
Analyzed wave	8	5	4	6	1
Learned wave1	8	6	4	6	5
Learned wave1	8	2	4	6	9

Table 6.4 Analyzed wave × Learned waves1

Wave's name	Position1	Position2	Position3	Position4	Position5	Total
Analyzed wave	8	5	4	6	1	–
Learned wave1	8	6	4	6	5	–
Favorable evidence	1	0	1	1	0	3
Contrary evidence	0	1	0	0	4	5

equipment, movement of the patient, electric registers and individual variability of physician's visual analysis. Such imprecision can often include conflicting information or paracomplete data.

The majority of theories and techniques available are based on classical logic and so they cannot handle adequately such sets of information, at least directly.

Table 6.5 Analyzed wave × Learned waves2

Wave's name	Position1	Position2	position3	Position4	Position5	Total
Analyzed wave	8	5	4	6	1	–
Learned wave2	8	2	4	6	9	–
Favorable evidence	1	0	1	1	0	3
Contrary evidence	0	3	0	0	8	11

Table 6.6 Normalized values: wave1 and learned waves2

Case	μ	λ	Normalization FE	Normalization CD
Analyzed wave × Learned wave1	3	5	0.6	1
Analyzed wave × Learned wave2	3	11	0.6	2.2

So, as PANN is based on logics $E\tau$, it can deal with uncertainty, inconsistent and paracomplete data. We believe that other types of annotated logics can be also used as foundations for PANNs.

Paraconsistent Artificial Neural Networks are a new kind of ANN based on annotated evidential logics $E\tau$, which can manipulate impreciseness, inconsistency and paracompleteness internally without trivialization. Its composition as a two-sorted language makes this property become critical, when complexity of hardware implementation of paraconsistent logical circuit and computational implementation are the main concerns. PANN is useful to the problems in data analysis, expert system, speech recognition, etc. Abe et al. [12] used PANNs to deal with Alzheimer's disease.

6.5 Automation and Robotics

Annotated logics $E\tau$ can be also used for automation and robotics. The *paraconsistent logical controller* (Paracontrol) is an electric-electronic realization of the para-analyzer algorithm, which is basically an electrical circuit, treating logical signals in $E\tau$; see Abe and Da Silva Filho [11, 69].

Such a circuit compares logical values and determines domains of a state lattice corresponding to output value. Favorable evidence and contrary evidence degrees are represented by voltage. Certainty and uncertainty degrees are determined by an analogue of operational amplifiers.

The Paracontrol comprises both analog and digital systems and it can be externally adjusted by applying positive and negative voltages. The Paracontrol was tested in real-life experiments with an autonomous mobile robot *Emmy*, whose favorable/contrary evidences coincide with the values of ultrasonic sensors and distances are represented by continuous values of voltage; see Da Silva Filho and Abe [71].[1]

[1] The name Emmy originates from the mathematician Amalie Emmy Nöwther (1882–1935).

Emmy consists of a circular mobile platform of aluminum 30 cm in diameter and 60 cm high. While moving in a non-structured environment, Emmy gets information about the presence/absence of obstacles using a sonar system called *Parasoninc*. Parasonic is able to detect obstacles in a robot's path by transforming the distance to the obstacle into electrical signal of continuous voltage ranging 0–5 V. Parasonic is basically composed of two ultrasonic sensors of type POLAROID 6500 controlled by an 8051 micro controller. The 8051 is programmed to carry out synchronization between the measurements of the two sensors and the transformation of the distance into electrical voltage.

Emmy uses the paracontrol system to negotiate traffic in non-constructed environments avoiding collisions with human beings, objects, walls, tables, etc. The form of reception of information on the obstacles is non-contact in nature, and is the method used to obtain and to treat ultra-sonic signals from optical sensors in order to avoid collisions. The circuit of Paracontrol is shown in Fig. 6.5.

The system of Emmy's control comprises Parasonic, Paracontrol and supporting circuits as in Fig. 6.6.

The Emmy robot control system has the following components (see Fig. 6.7):

- Ultra-Sonic Sensors: The two sensors detect the distance between the robot and the object through the emission of a pulse train of ultra-sonic sound waves and the reception of the signal (echo).
- Signal Treatment: The treatment of the captured signals is made through the Para-sonic. The microprocessor is programmed to transform the time

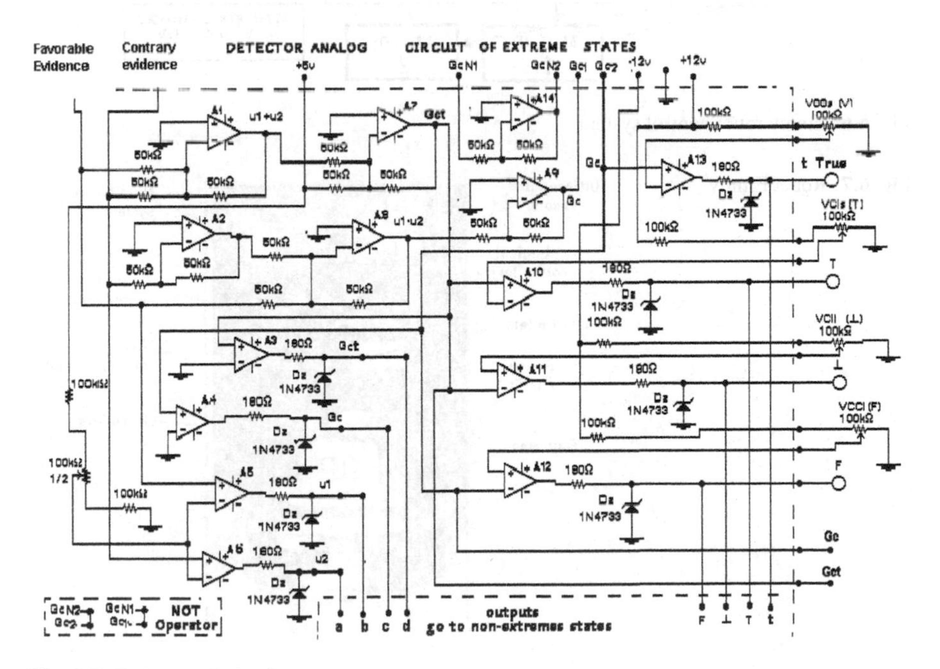

Fig. 6.5 Paracontrol circuit

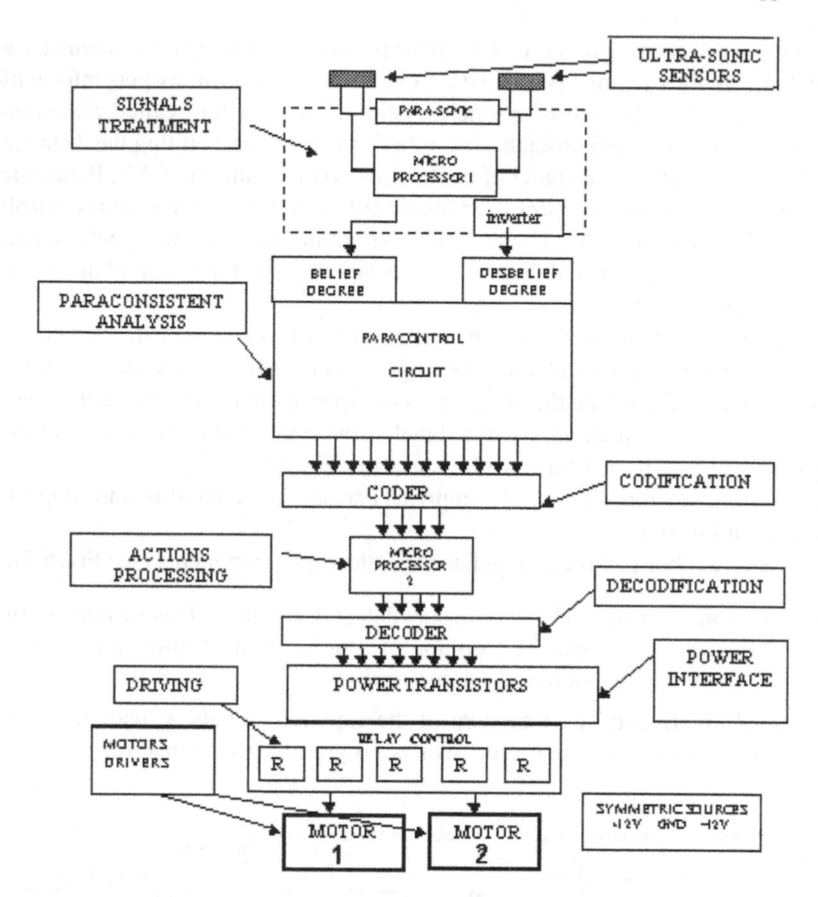

Fig. 6.6 Emmy robot control system

Fig. 6.7 Robot Emmy

elapsed between the emission of the signal and the reception of the echo in an electric signal in the range 0 to 5 V for belief degree, and from 5 to 0 V for disbelief degree. The width of each voltage is proportional to the time that has elapsed between the emission of a pulse train and its reception by sensors.

- Paraconsistent Analysis: The paraconsistent control logic makes a logical analysis of the signals according to the logic $E\tau$.
- Codification: The coder circuit changes the binary word of 12 digits to a code of 4 digits to be processed by the personal computer.
- Action Processing: The microprocessor is programmed conveniently to work the relay in sequences to establish actions for the robot.
- Decodification: The circuit decoder changes in the 4-digit binary word in order to charge the relay in the programmed paths.
- Power Interface: The power interface circuit is composed of transistors that amplify the signals making possible the relay's control by digital signals.
- Driving: The relays ON and OFF control the motors M_1 and M_2 according to the decoded binary word.
- Motor Drives: The motors M_1 and M_2 move the robot according to the relay control sequence.
- Sources: The robot Emmy is supplied by two batteries forming a symmetrical source of tension to generate ± 12 V DC.

As the project is built in hardware besides the paracontrol, it was necessary to install components for supporting circuits allowing the resulting signals of the paraconsistent analysis to be addressed and indirectly transformed into action.

In this first prototype of the robot Emmy it was necessary to incorporate an encoder and a decoder such that the referring signals to the logical states resulting from the paraconsistent analysis was processed by a microprocessor of 4 inputs and 4 outputs.

Later, we also developed the robot called *Emmy II*; see Abe et al. [15]. The platform used to assemble the Emmy II robot measures approximately 23 cm high and 25 cm diameter (circular format). The main components of Emmy II are an 8051 microcontroller, two ultrasonic sensors and two DC motors. Figure 6.8 shows the basic structure of Emmy II.

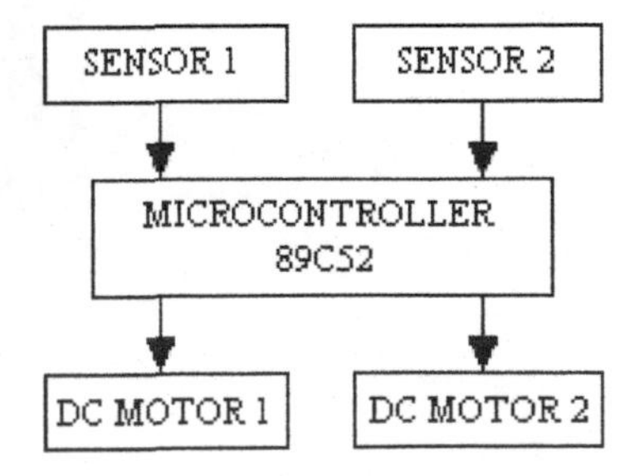

Fig. 6.8 Basic structure of Emmy II

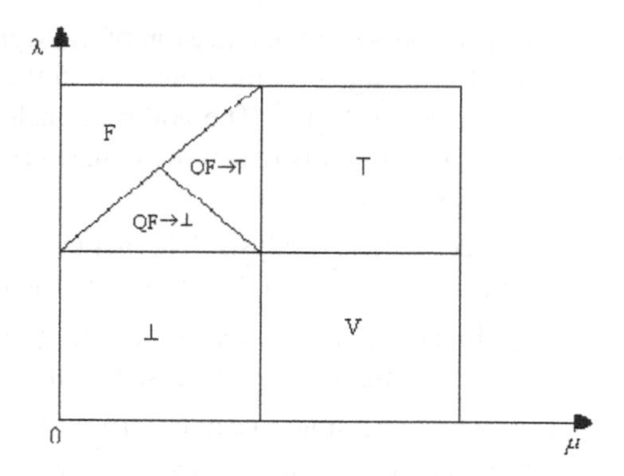

Fig. 6.9 Logical output lattice

In Emmy II, the signals from the 2 sensors are used to determine the favorable evidence degree μ and the contrary degree evidence degree λ regarding the proposition "There is no obstacle in front of the robot".

Then, the Paracontrol, recorded in the internal memory of the microcontroller is used to determine the robot movements. Also, the microcontroller is responsible for applying power to the DC motors.

The decision-making of the Emmy II regarding the movements to undertake is according to the decision state lattice shown in Fig. 6.9.

Each robot movement lasts approximately 0.4 s. The source circuitry supplies 5, 12 and 16.8 V DC to the other robot Emmy II circuits. Figures 6.10 and 6.11 are pictures of Emmy II.

The "Polaroid 6500 Series Sonar Ranging Module" was used in the Emmy II robot. This device has three inputs (Vcc, Gnd and Init) and one output (Echo). When INIT is taken high by the microcontroller, the sonar ranging module transmits 16 pulses at 49.4 kHz.

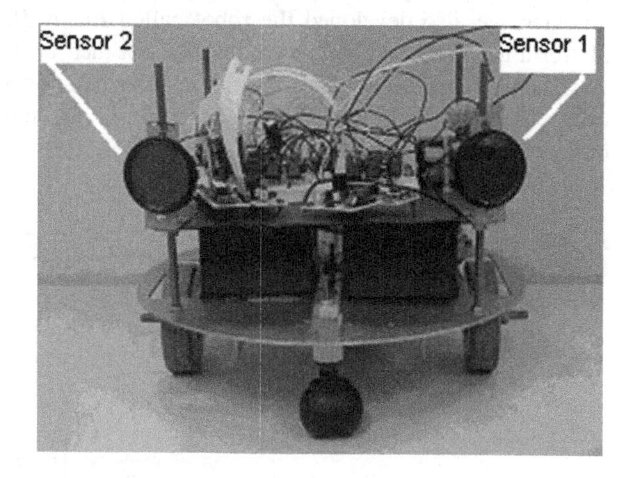

Fig. 6.10 Front view of Emmy II

Fig. 6.11 Inferior view of Emmy II

After receiving the echo pulses, which cause the ECHO output to go high, the sonar ranging module sends an ECHO signal to the microcontroller. Then, with the time interval between INIT sending and ECHO receiving, the microcontroller is able to determine the distance between the robot and the obstacle.

The microcontroller 89C52 from the 8051 family is responsible to control the Emmy II robot. Its input/output port 1 is used to send and receive signals to and from the Sensor Circuitry and Power Circuitry. Buffers are used in the interface between the microcontroller and the other circuits.

Logical states and their actions are given in Table 6.7.

Figure 6.12 shows the source circuitry of the electric scheme of Emmy II.

Figure 6.13 shows the sensor circuit of Emmy II.

The microcontroller I/O Port 1 has 8 pins. The function of each pin is as follows:

- Pin 0: INIT of sonar ranging module 1 (S_1)
- Pin 1: ECHO of sonar ranging module 1 (S_1)
- Pin 2: INIT of sonar ranging module 2 (S_2)

Table 6.7 Logical states and action

Symbol	State	Action
V	True	Robot goes ahead
F	False	Robot goes back
$\perp$	Paracomplete	Robot turns right
T	Inconsistent	Robot turns left
$QF \rightarrow \perp$	Quasi-false tending to paracomplete	Robot turns right
$QF \rightarrow T$	Quasi-true tending to inconsistent	Robot turns left

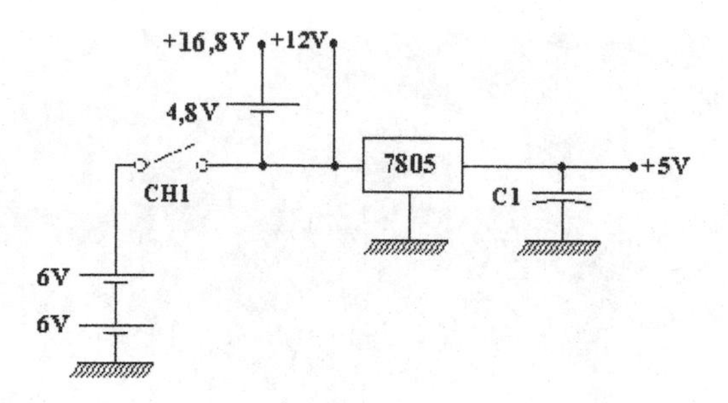

Fig. 6.12 Source circuitry

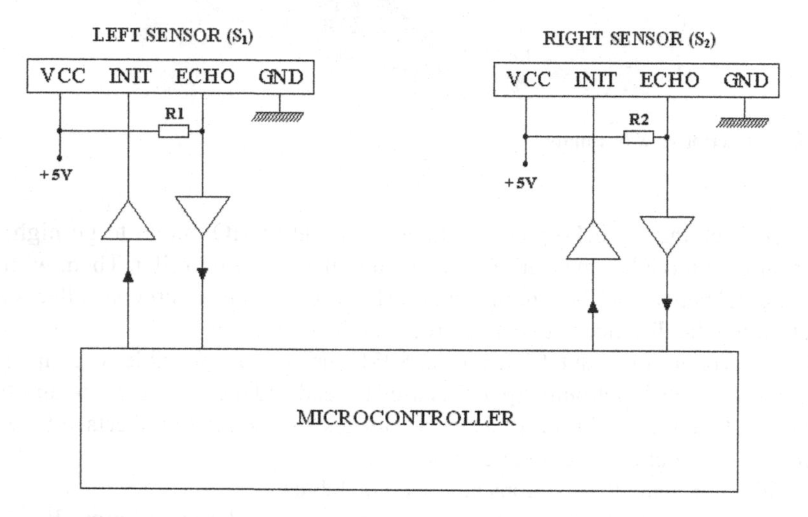

Fig. 6.13 Sensor circuit

- Pin 3: ECHO of sonar ranging module 2 (S_2)
- Pin 4: When it is taken high (+5 V), DC motor 1 has power supplied
- Pin 5: When it is taken high (+5 V), DC motor 2 has power supplied
- Pin 6: When it is taken low (0 V), DC motor 1 spins around forward while pin 4 is taken high (+5 V). When it is taken high (+5 V), DC motor 1 spins around backward while pin 4 is taken high (+5 V).
- Pin 7: When it is taken low (0 V), DC motor 2 spins around forward while pin 5 is taken high (+5 V). When it is taken high (+5 V), DC motor 2 spins around backward while pin 5 is taken high (+5 V).

Two DC motors supplied by 12 V DC are responsible for Emmy II robot movements. The Paracontrol, through the microcontroller, determines which DC motor must be supplied and in which direction it must spin.

Basically, the power interface circuitry is comprised of power field effect transistors-MOSFETs.

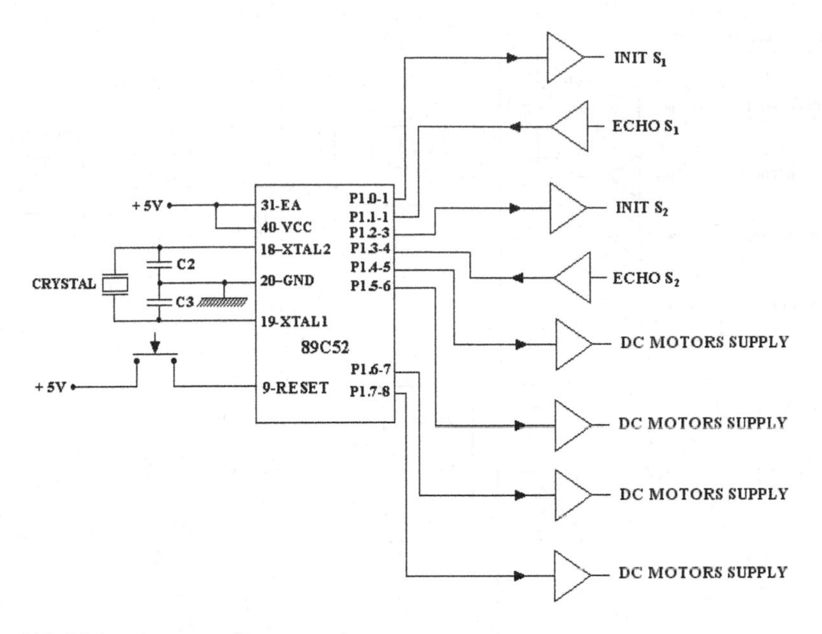

Fig. 6.14 Main microcontroller connections

Fig. 6.15 Test environment

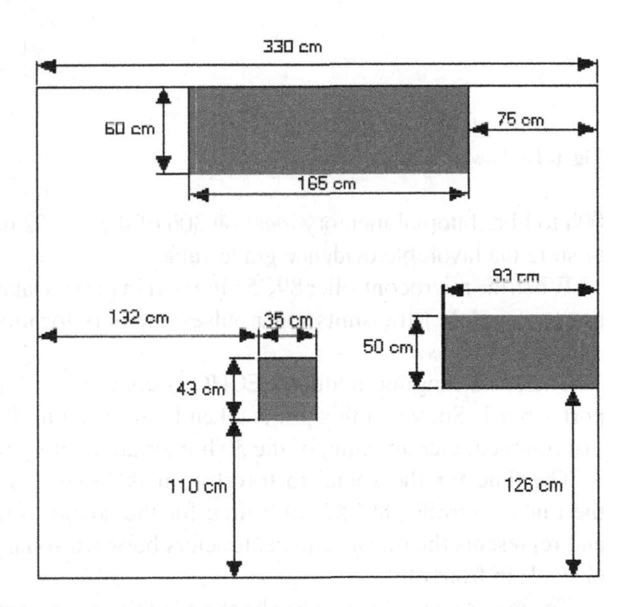

Figure 6.14 shows the main microcontroller connections of Emmy II.

Figure 6.15 represents a test environment.

Figure 6.16 represents the power interface circuitry.

The microcontroller 89C52 internal memory can store numbers of 8 bits. If we represent these numbers in hexadecimal, it means that we can store numbers from

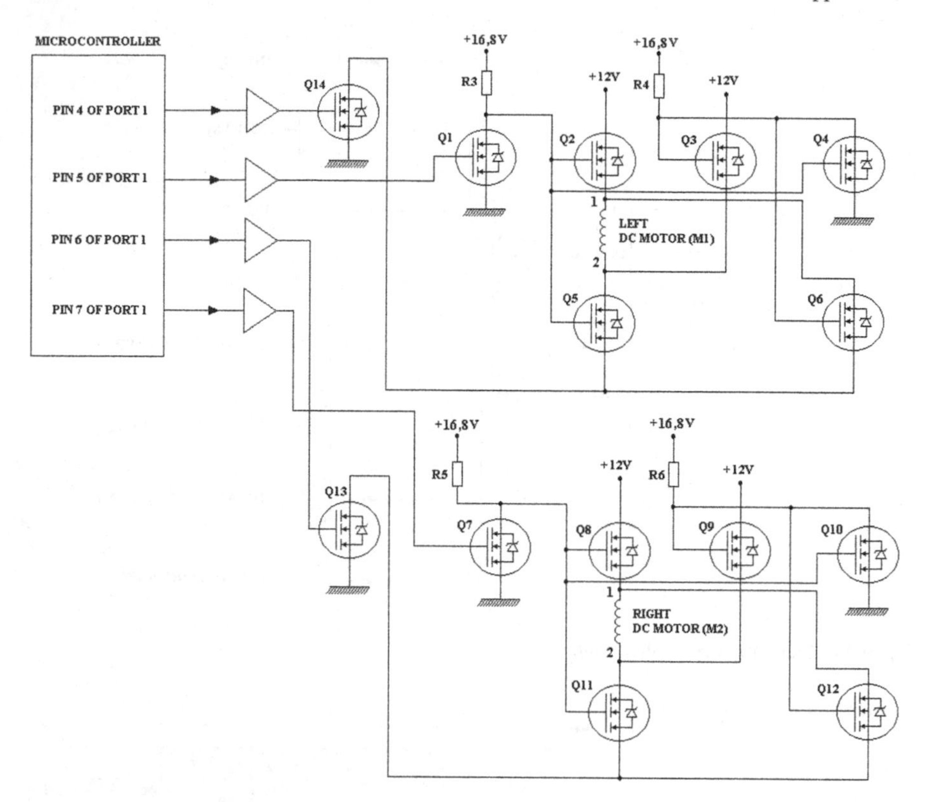

Fig. 6.16 Power interface circuitry

00h to FFh. Internal memory location 30h of the 89C52 microcontroller was chosen to store the favorable evidence grade value.

When the microcontroller 89C52 I/O port in pin 0 is taken high, meaning that sonar ranging module 1 transmits sonar pulses, memory location 30h starts to increase by 1 unit per 118 μs.

The sonar ranging module 1 ECHO is connected to microcontroller 89C52 I/O port 1 pin 1. So, when this pin is taken high, meaning that sonar echo pulses have just returned, incrementing of the 30 h internal memory position ceases.

The time for the sound to travel 4 cm is 118 μs. Hence, The value stored in the microcontroller 89C52 30 h time for the sound to travel 4 is multiplied by 2 and represents the distance in centimeters between sonar ranging module 1 and the obstacle in front of it.

The maximum value that can be stored in the microcontroller 89C52 30 h internal memory position is 3Fh (to allow arithmetic calculation). Thus, the sonar ranging module 1 can measure distances between 00h and 3Fh, or 0 cm and 126 cm.

Therefore, 00h stored in the microcontroller 89C52 30h internal memory position means that the favorable evidence grade value (μ) on the proposition "There is no obstacle in front of the robot" is 0.

3Fh stored in the microcontroller 89C52 30h internal memory position means that the intermediate favorable evidence grade value (μ) on the proposition "There is no obstacle in front of the robot" is 1.

In the same way, intermediate values between 00h and 3Fh stored in the microcontroller 89C52 30h internal memory position means intermediate favorable evidence grade values (μ) on the proposition "There is no obstacle in front of the robot".

When the microcontroller 89C52 I/O port 1 pin 2 is taken high, meaning that sonar ranging module 2 transmits sonar pulses, 31h memory position starts to be increased by 1 unit per 118 μs.

The sonar ranging module 2 ECHO is connected to microcontroller 89C52 I/O port 1 pin 3. So, when this pin is taken high, meaning that sonar echo pulses have just returned, incrementing of the microcontroller 89C52 internal position 31h ceases.

The time for the sound to travel 4 cm is 118 μs. Hence, the value stored in the microcontroller 89C52 31h internal memory position multiplied by 2 represents the distance in centimeters between sonar ranging module 2 an the obstacle in front of it.

The maximum value that can be stored in the microcontroller 89C52 31h internal memory position is 3Fh. Then, the sonar ranging module 2 can measure distances between 00h and 3Fh, or 0 cm and 126 cm.

The microcontroller 89C52 31h internal memory position is complemented. Thus, it represents the contrary evidence grade value (λ) on the proposition "There is no obstacle in front of the robot".

Consequently, 00h stored in the microcontroller 89C52 31h internal memory position means that the contrary evidence grade value (λ) on the proposition: "There is no obstacle in front of the robot" is 0.

And 4Fh stored in the microcontroller 89C52 31h internal memory position means that the contrary evidence rate value (λ) on the proposition"There is no obstacle in front of the robot" is 1.

In the same way, intermediate values between 00h and 3Fh stored in the microcontroller 89C52 31h internal memory position means intermediate contrary evidence grade value (λ) on the proposition "There is no obstacle in front of the robot".

We also developed the autonomous mobile robot called *Emmy III*; see Desiderto and De Oliveira [73]. The aim of the Emmy III is to be able to move from an origin point to an end point, both predetermined, in a non-structured environment. The first prototype of Emmy III is composed of a planning system and a mechanical construction.

The planning system considers an environment to be divided into cells. The first version considers all cells to be free. Firstly, the robot must be able to move from a point to another without encountering any obstacle. This environment is divided into cells as in Elfes [78] and a planning system gives the sequence of cells the robot must travel in order to reach the end cell.

Then, it asks for the initial point and the target point. After that a sequence of movements is given. Also a sequence of pulses is sent to the stepping motors that are responsible for moving the physical platform of the robot. In this manner, the robot moves from the initial point to the target point. The physical construction of

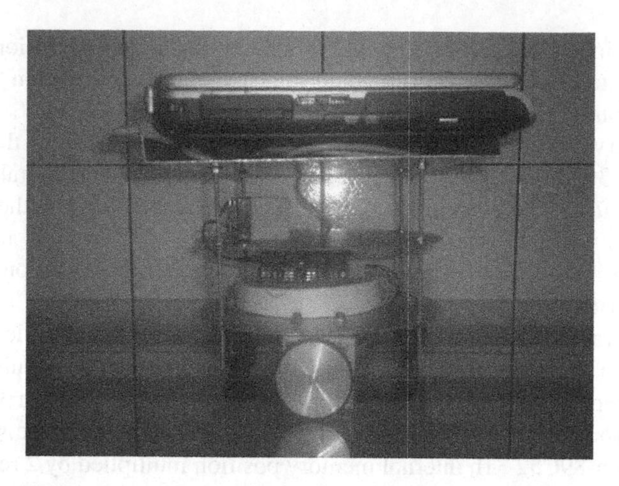

Fig. 6.17 Emmy III

the first prototype of the Emmy III robot is basically composed of a circular platform of approximately 286 mm diameter and two step motors.

Figure 6.17 shows the Emmy III first prototype.

The planning system is recorded in a notebook, and the communication between the notebook and the physical construction is made through the parallel port. A potency driver is responsible for getting the pulses from the note book and sending them to the stepping motors.

Like the first prototype, the second prototype of Emmy III is basically composed of a planning system and a mechanical structure. The planning system can be recorded in any personal computer and the communication between the personal computer and the mechanical construction is done through a USB port.

The planning system considers the environment around the robot to be divided into cells. So, it is necessary to inform the planning system which the cell the robot is in and likewise for the target cell.

Then, the planning system gives a sequence of cells that the robot must follow to go from the origin cell to the target cell. The planning system considers all cells to be free.

The mechanical construction is basically composed of a steel structure, two DC motors and three wheels. Each motor has a wheel field in its axis and there is a free wheel.

Figure 6.18 shows the mechanical structure.

There is electronic circuitry on the steel structure. The main device of the electronic circuit is the microcontroller PIC18F4550 that is responsible for receiving the schedule from the planning system and activating the DC motors. There is also a potency driver between the microcontroller and the DC motors.

Recently, Emmy II was implemented in a new hardware structure. Basically, the robot there is similar to Emmy II, but with a modern hardware structure to receive upgrades in it functionalities; see Torres and Reis [152].

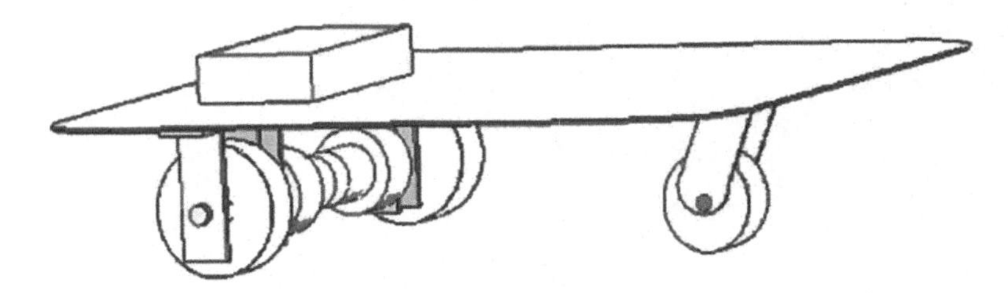

Fig. 6.18 Basic structure of Emmy III

For thousands of years, humans have researched to find logical structures to 'imitate' human thinking and actions, in order to implement robots and automated processing. As the majority of automated processing and robotics are based on classical logic, in a certain sense their actions are predictable; all commands are of the type yes/no, because classical logic is two-valued.

So, robots built with classical logic are too limited in that they cannot handle conflicting and imprecise information. So, we need more flexible logical systems for robots for them to perform more complex tasks in a smooth way without becoming involved with conflicts, imprecision and paracompleteness.

This means that a new class of non-classical logics should be developed and incorporated into some interesting autonomous mobile robots, which can deal with imprecise, inconsistent and paracomplete data. As we saw, annotated logics, in particular, $E\tau$, can be used to develop such robots. The fact is not surprising because annotated logics can naturally capture human reasoning.

The applications to robotics considered here show the power and the beauty of paraconsistent systems. They have provided other ways than usual electronic theory, opening applications of non-classical logics in order to overcome situations not covered by other systems or techniques. All the topics treated in this section are being improved as well as other projects being undertaken in a variety of themes.

Chapter 7
Conclusions

Abstract This chapter gives some conclusions with the summary of the book. It is possible to conclude that annotated logics are very interesting theoretically as well as practically. However, there are some future problems to be worked out.

7.1 Summary

The origin of annotated logics was the area of logic programming, capable of reasoning about incomplete and inconsistent information. Later, some people started foundational work to study annotated logics as non-classical logics. This book compiled major results in the study of annotated logics.

We can formalize propositional and predicate annotated logics with proof and model theory. Annotated logics can be also applied to set theory. There are some variants of the original formulation of annotated logics and rival logical systems in the literature. Chapters 2–5 were concerned with foundational study.

Annotated logics have various applications as seen in Chap. 6. We reviewed some promising applications, i.e., paraconsistent logic programming, generalized annotated logic programming, knowledge representation, neural computing, and automation and robotics. The usefulness of annotated logics is due to the fact they are paraconsistent and paracomplete. It should be also pointed out that annotated logics have intuitive semantics and computational proof theory.

The fact that annotated logics have both theoretical foundations and various applications reveals the advantage of annotated logics over other existing logical systems. Indeed a number of logics were proposed to model incomplete and inconsistent information, but most of them are far from being satisfactory. In this sense, annotated logics can be considered as more attractive than other current logical systems.

J.M. Abe et al., *Introduction to Annotated Logics*,
Intelligent Systems Reference Library 88, DOI 10.1007/978-3-319-17912-4_7

7.2 Future Problems

There are some future problems both in theoretical and practical aspects. We discuss these topics briefly, and we shall tackle them on other occasions and suggest them to readers who want to study annotated logics.

We believe that foundations of annotated logics have been basically established. Other interesting problems in foundations are as follows. First, cut-free sequent calculus should be investigated. We presented both Hilbert and natural deduction systems, but they are not suited to automate reasoning in annotated logics.

In this regard, cut-free sequent calculus is very important from the point of view of the cut-elimination theorem and of tableau calculus, resolution and logic programming.

Since annotated predicate logics use two types of variables, it may be able to advance a two-sorted interpretation of annotated logics. The interpretation has merit in that techniques of first-order model theory can be applied. It is also useful in that we can consider many-sorted theorem-proving.

For model theoretic aspects, there are several extensions. Formulas generated by the equality symbol can also be 'annotated', although equality formulas in $Q\tau$ are interpreted classically. We can also develop versions of $Q\tau$ without equality, and study their model theory.

It is interesting to develop some versions of annotated higher-order logic (e.g. type theory and set theory). These logics appear to be suitable for paraconsistent mathematics; see Mortensen [122]. Several results on annotated set theory have been reported in Chap. 4.

There seem to be other types of semantics for annotated logics. We should investigate the *Boolean-valued model*, because it is very flexible and is applied to various (non-classical) logics; see Bell [47]. There are surely other possibilities of formulating algebraic semantics.

One of the most interesting topics is to prove completeness for annotated logics $P\tau$ based on infinite lattices. The completeness results obtained in this book concern annotated logics based on finite lattice. The problem lies in the fact we need an infinitary rule for infinite lattices. One idea to prove completeness in this context is to use α*-models* due to da Costa [60].

We described annotated modal logics in Chap. 4. It is interesting to develop other modal extensions of annotated logics. Modal logic of knowledge and beliefs is an important research topic in AI and computer science; see Halpern and Moses [88, 89] and Fagin et al. [80].

Therefore, annotated modal logics of knowledge and belief are needed for AI applications. These annotated logics can be viewed as improvements of Hintikka's work in Hintikka [90]. It is important to work out annotated temporal and deontic logics.

Annotated logics are originally based on (positive) classical logic, but we would be able to use other bases. For example, intuitionistic logic and relevance logic can be used as the base. Generally, we can consider any *substructural logic* as the base logic; see Restall [141].

However, such annotated logics may be able to provide a new way of formalizing paraconsistent and paracomplete logics. However, the problem is not easy, because we need to work out complicated versions of Kripke semantics.

For probabilistic reasoning, annotated logics can be extended. There are several theories of probability theory and probabilistic logics. In any case, we should consider a different algebraic structure for atomic formulas in annotated logics, since a complete lattice is not always suitable for probabilistic reasoning. Ng and Subrahmanian [134] developed *probabilistic logic programming* that is syntactically similar to annotated logics but in which the truth-values are interpreted probabilistically.

Related to this, evidential annotated logics $E\tau$ should also be elaborated to address the connection with *Dempster-Shafer theory* (DST); see Dempster [72] and Shafer [146].

If annotated logics are a class of paraconsistent logics, formalizing *naive set theory* based on annotated logics is an important topic. Natural deduction formulation is useful for this purpose, since the normal form theorem for natural deduction leads us to the non-triviality of annotated naive set theory.

Annotated logics can be the underlying logic of an annotated set theory that is more powerful than usual fuzzy set theory. On the other hand, it also seems possible to develop a kind of higher-order annotated logic (linked to the analysis of usual languages). In the future, these subjects will be very important. On the other hand, annotated set theory could, in addition, be a type of paracomplete logical system. Paraconsistency and paracompleteness are strictly correlated.[1]

We turn to other avenues of applications of annotated logics, since we only described current ones in this book. As an application to knowledge representation, F-logic can be extended by combining it with annotated logics. This leads to a knowledge representation language which can deal both with incomplete and inconsistent knowledge.

Annotated logics can serve as the foundation for non-monotonic reasoning. In particular, we can focus on the aspects of inconsistency in common-sense reasoning. For this purpose, we have to integrate existing non-monotonic logics and annotated logics.

For applications to robotics, it is expected to realize a more sophisticated way of control based annotated logics. To develop human-like robots, we need to incorporate inference engines for humans by means of annotated logics.

Finally, we have to seek other application domains for annotated logics. For example, applications to signal processing, image processing, and decision-making are interesting application areas.

We believe that annotated logics can be also applied to *quantum computing*; see Deutsch [74]. This is because the formalization of logic-based quantum computing is closely related to the concepts of paraconsistency and paracompleteness; see Akama [22].

---

[1] The issue was suggested by Prof. da Costa.

References

1. Abe, J.M.: On the foundations of annotated logics (in Portuguese). Ph.D. thesis, University of São Paulo, Brazil (1992)
2. Abe, J.M.: On annotated modal logics. Mathematica Japonica **40**, 553–560 (1994)
3. Abe, J.M.: Some recent applications of paraconsistent systems to AI. Logique et Analyse **157**, 83–96 (1997)
4. Abe, J.M.: Curry algebra $P\tau$. Log. Anal. **161–163**, 5–15 (1998)
5. Abe, J.M.: Paraconsistent artificial neural networks; an introduction. In: Carbonell, J.G., Siekmann, J. (eds.) Lecture Notes in Artificial Intelligence, vol. 3214, pp. 942–948. Springer, Heidelberg (2004)
6. Abe, J.M., Akama, S.: A logical system for reasoning with fuzziness and inconsistencies in distributed systems. In: Proceedings of IASTED, Honolulu, Hawaii, pp. 221–225 (1999)
7. Abe, J.M., Akama, S.: Annotated logics $Q\tau$and ultraproduct. Logique et Analyse **160**, 335–343 (1997) (published in 2000)
8. Abe, J.M., Akama, S.: On some aspects of decidability of annotated systems. In: Arabnia, H.R. (ed.) Proceedings of the International Conference on Artificial Intelligence, vol. II, pp. 789–795. CREA Press (2001)
9. Abe, J.M., Akama, S.: Annotated temporal logics $\Delta\tau$. In: Advances in Artificial Intelligence: Proceedings of IBERAIA-SBIA 2000. LNCS, vol. 1952, pp. 217–226. Springer, Berlin (2000)
10. Abe, J.M., Akama, S., Nakamatsu, K.: Monadic curry algebras $Q\tau$. In: Knowledge-Based Intelligent Information and Engineering Systems: Proceedings of KES 2007—WIRN 2007, Part II. Lecture Notes on Artificial Intelligence, vol. 4693, pp. 893–900 (2007)
11. Abe, J.M., Da Silva Filho, J.I.: Manipulating conflicts and uncertainties in robotics. Multi-Valued Log. Soft Comput. **9**, 147–169 (2003)
12. Abe, J.M., Lopes, H.F.S., Anghinah, R.: Paraconsistent artificial neural network and Alzheimer disease: a preliminary study. Dementia and Neuropsychologia **3**, 241–247 (2007)
13. Abe, J.M., Nakamatsu, K., Akama, S.: Two applications of paraconsistent logical controller. In: Tsihrintzis, G.A., Virvou, M., Howlett, R.J., Jain, L.C. (eds.) New Directions in Intelligent Interactive Multimedia, pp. 249–254. Springer, Berlin (2008)
14. Abe, J.M., Nakamatsu, K., Akama, S.: An algebraic version of the monadic system C_1. In: Proceedings of KES-IDT2009 (2009)
15. Abe, J.M., Torre, C.R., Torre, G.L., Nakamatsu, K., Kondo, M.: Intelligent paraconsistent logic controller and autonomous mobile robot Emmy II. Lecture Notes in Computer Science, vol. 4252, pp. 851–857. Springer, Heidelberg (2006)
16. Akama, S.: Resolution in constructivism. Logique et Analyse **120**, 385–399 (1987)
17. Akama, S.: Constructive predicate logic with strong negation and model theory. Notre Dame J. Form. Log. **29**, 18–27 (1988)

J.M. Abe et al., *Introduction to Annotated Logics*,
Intelligent Systems Reference Library 88, DOI 10.1007/978-3-319-17912-4

18. Akama, S.: On the proof method for constructive falsity. Zeitschrift für mathematische Logik und Grundlagen der Mathematik **34**, 385–392 (1988)
19. Akama, S.: Subformula semantics for strong negation systems. J. Philos. Log. **19**, 217–226 (1990)
20. Akama, S.: Constructive falsity: foundations and their applications to computer science. Ph.D. thesis, Keio University, Yokohama, Japan (1990)
21. Akama, S.: Nelson's paraconsistent logics. Log. Log. Philos. **7**, 101–115 (1999)
22. Akama, S.: Elements of Quantum Computing: History. Theories and Engineering Applications. Springer, Heidelberg (2015)
23. Akama, S., Abe, J.M.: Many-valued and annotated modal logics. In: Proceedings of the 28th International Symposium on Multiple-Valued Logic, Fukuoka, pp. 114–119 (1998)
24. Akama, S., Abe, J.M.: Natural deduction and general annotated logics. In: Proceedings of the 1st International Workshop on Labelled Deduction, Freiburg (1998)
25. Akama, S., Abe, J.M.: Fuzzy annotated logics. In: Proceedings of IPMU'2000, Madrid, Spain, pp. 504–508 (2000)
26. Akama, S., Abe, J.M.: The degree of inconsistency in paraconsistent logics. In: Abe, J.M., da Silva Filho, J.I. (eds.) Logic, Artificial Intelligence and Robotics, pp. 13–23. IOS Press, Amsterdam (2001)
27. Akama, S., Abe, J.M.: Paraconsistent logics viewed as a foundation for data warehouse. In: Abe, J.M., da Silva Filho, J.I. (eds.) Advances in Logic, Artificial Intelligence and Robotics, pp. 96–103. IOS Press, Sao Paulo (2002)
28. Akama, S., Abe, J.M., Murai, T.: On the relation of fuzzy and annotated logics. In: Proceedings of ASC'2003, Banff, Canada, pp. 46–51 (2003)
29. Akama, S., Abe, J.M., Murai, T.: A tableau formulation of annotated logics. In: Cialdea Mayer, M., Pirri, F. (eds.) Proceedings of TABLEAUX'2003, Rome, Italy, pp. 1–13 (2003)
30. Akama, S., Abe, J.M., Nakamatsu, K.: Constructive discursive logic with strong negation. Logique et Analyse **215**, 395–408 (2011)
31. Akama, S., Nakamatsu, K., Abe, J.M.: A natural deduction system for annotated predicate logic. In: Knowledge-Based Intelligent Information and Engineering Systems: Proceedings of KES 2007—WIRN 2007, Part II. Lecture Notes on Artificial Intelligence, vol. 4693, pp. 861–868. Springer, Berlin (2007)
32. Almukdad, A., Nelson, D.: Constructible falsity and inexact predicates. J. Symb. Log. **49**, 231–233 (1984)
33. Andreka, H., Nemeti, I.: The generalized completeness of Horn logic as a programming language. Acta Cybernetica **4**, 3–10 (1978)
34. Anderson, A., Belnap, N.: Entailment: The Logic of Relevance and Necessity, vol. I. Princeton University Press, Princeton (1976)
35. Anderson, A., Belnap, N., Dunn, J.: Entailment: The Logic of Relevance and Necessity, vol. II. Princeton University Press, Princeton (1992)
36. Arieli, O., Avron, A.: Reasoning with logical bilattices. J. Log. Lang. Inf. **5**, 25–63 (1996)
37. Arieli, O., Avron, A.: The value of four values. Artif. Intell. **102**, 97–141 (1998)
38. Arruda, A.I.: On the imaginary logic of N.A. Vasil'ev. In: Arruda, A., da Costa, N., Chuaqui, R. (eds.) Non-Classical Logic, Model Theory and Computability, pp. 3–24. North-Holland, Amsterdam (1977)
39. Arruda, A.I.: A Survey of Paraconsistent Logic. In: Arruda, A., da Costa, N., Chuaqui, R. (eds.) Mathematical Logic in Latin America, pp. 1–41. North-Holland, Amsterdam (1980)
40. Asenjo, F.G.: A calculus of antinomies. Notre Dame J. Form. Log. **7**, 103–105 (1966)
41. Avila, B.C., Abe, J.M., Prado, J.P.A: ParaLog-e: a paraconsistent evidential logic programming language. In: Proceedings of the 17th International Conference on the Chilean Computer Society, Valparaiso, pp. 2–8. IEEE Computer Society Press, California (1997)
42. Barros, C.M., da Costa, N.C.A., Abe, J.M.: Tópico de teoria dos sistemas ordenados: vol II, sistemas de Curry, (in Portuguese). Coleção Documentos, Série Lógica e Teoria da Ciência, IEA-USP **20** (1995)

43. Batens, D.: Dynamic dialectical logics. In: Priest, G., Routley, R., Norman, J. (eds.) Paraconsistent Logic: Essay on the Inconsistent, pp 187–217. Philosophia Verlag, München (1989)
44. Batens, D.: Inconsistency-adaptive logics and the foundation of non-monotonic logics. Logique et Analyse **145**, 57–94 (1994)
45. Batens, D.: A general characterization of adaptive logics. Logique et Analyse **173–175**, 45–68 (2001)
46. Batens, D., Mortensen, C., Priest, G., Van Bendegem, J.-P. (eds.): Frontiers of Paraconsistent Logic. Research Studies Press, Baldock (2000)
47. Bell, J.L.: Boolean-Valued Models and Independence Proofs in Set Theory. Clarendon Press, Oxford (1985)
48. Bell, J.L., Somson, A.B.: Models and Ultraproducts: An Introduction. North-Holland, Amsterdam (1977)
49. Belnap, N.D.: A useful four-valued logic. In: Dunn, J.M., Epstein, G. (eds.) Modern Uses of Multi-Valued Logic, pp. 8–37. Reidel, Dordrecht (1977)
50. Belnap, N.D.: How a computer should think. In: Ryle, G. (ed.) Contemporary Aspects of Philosophy, pp. 30–55. Oriel Press, Stocksfield (1977)
51. Besnard, P., Lang, J.: Graded paraconsistency-reasoning with inconsistent and uncertain knowledge. In: Batens, D., Mortensen, C., Priest, G., Van Bendegem, J.-P. (eds.) Frontiers of Paraconsistent Logic, pp. 75–94. Research Studies Press, Baldock (2000)
52. Beziau, J.-Y., Carnielli, W., Gabbay, D. (eds.): Handbook of Paraconsistency. College Publication, London (2007)
53. Blair, H.A., Subrahmanian, V.S.: Paraconsistent logic programming. Theor. Comput. Sci. **68**, 135–154 (1989)
54. Blok, W.J., Pigozzi, D.: Algebraizable logics. Mem. AMS (1989)
55. Carnielli, W.A., Coniglio, M.E., D'Ottaviano, I.M. (eds.): Paraconsistency: The Logical Way to the Inconsistent. Marcel Dekker, New York (2002)
56. Carnielli, W.A., Coniglio, M.E., Marcos, J.: Logics of formal inconsistency. In: Gabbay, D., Guenthner, F. (eds.) Handbook of Philosophical Logic, vol. 14, 2nd edn., pp. 1–93. Springer, Heidelberg (2007)
57. Carnielli, W.A., Marcos, J.: Tableau systems for logics of formal inconsistency. In: Abrabnia, H.R. (ed.), Proceedings of the 2001 International Conference on Artificial Intelligence, vol. II, pp. 848–852. CSREA Press, Las Vegas (2001)
58. Cohen, P.: Set Theory and the Continuum Hypothesis. Benjamin, New York (1963)
59. Curry, H.B.: Foundations of Mathematical Logic. Dover, New York (1977)
60. da Costa, N.C.A.: α-models and the system T and T^*. Notre Dame J. Form. Log. **14**, 443–454 (1974)
61. da Costa, N.C.A.: On the theory of inconsistent formal systems. Notre Dame J. Form. Log. **15**, 497–510 (1974)
62. da Costa, N.C.A., Abe, J.M., Subrahmanian, V.S.: Remarks on annotated logic. Zeitschrift für mathematische Logik und Grundlagen der Mathematik **37**, 561–570 (1991)
63. da Costa, N.C.A., Alves, E.H.: A semantical analysis of the calculi C_n. Notre Dame J. Form. Log. **18**, 621–630 (1977)
64. da Costa, N.C.A., Henschen, L.J., Lu, J.J., Subrahmanian, V.S.: Automatic theorem proving in paraconsistent logics: foundations and implementation. In: Proceedings of the 10th International Conference on Automated Deduction, pp. 72–86. Springer, Berlin (1990)
65. da Costa, N.C.A., Subrahmanian, VS.: Paraconsistent logic as a formalism for reasoning about inconsistent knowledge. Artif. Intell. Med. **1**, 167–174 (1989)
66. da Costa, N.C.A., Subrahmanian, V.S., Vago, C.: The paraconsistent logic PT. Zeitschrift für mathematische Logik und Grundlagen der Mathematik **37**, 139–148 (1991)
67. da Costa, N.C.A., Krause, D.: An inductive annotated logic. In: Carnielli, W., Coniglio, M., D'Ottaviano, I. (eds.) Paraconsistency: The Logical Way to the Inconsistent, pp. 213–225. Marcel Dekker, New York (2002)
68. da Costa, N., Prado, J., Abe, J.M., Avila, B., Rillo, M.: Paralog: Um Prolog paraconsistente baseado em Logica Anotada. Colecao Documentos, Serie Logica e Teoria da Ciencia, IEA-USP, vol. 18 (1995)

69. Da Silva Filho, J.I.: Métodos de interpretação da Lógica Paraconsistente Anotada com anotação com dois valores LPA2v com construção de Algoritmo e implementação de Circuitos Eletrônicos (in Portuguese). Doctor Thesis, University of São Paulo (1999)
70. Da Silva Filho, J.I., Abe, J.M.: Fundametos das Redes Neurais Paraconsistentes (in Portuguese). Editora Arte & Ciencia, São Paulo (2001)
71. Da Silva Filho, J.I., Abe, J.M.: Emmy: a paraconsistent autonomous mobile robot. In: Abe, J.M., Da Silva Filho, J.I. (eds.) Frontiers in Artificial Intelligence and its Applications, pp. 53–61. IOS Press, Amsterdam (2001)
72. Dempster, A.P.: Upper and lower probabilities induced by a multivalued mapping. Ann. Math. Stat. **38**, 325–339 (1967)
73. Desiderato, J.M.G., De Oliveira, E.N.: Primeiro Prototipo do Dobo Movel Autonoo Emmy III (in Portuguese), University of São Paulo (2006)
74. Deutsch, D.: Quantum theory, the Church-Turing principle and the universal quantum computer. Proc. R. Soc. A **400**, 97–117 (1985)
75. Dubois, D., Lang, J., Prade, H.: Possibilistic logic. In: Gabbay, D., Hogger, C., Robinson, J.A. (eds.) Handbook of Logic in Artificial Intelligence and Logic Programming, pp. 439–513. Oxford University Press, Oxford (1994)
76. Dubois, D., Konieczny, S., Prade, H.: Quasi-possibilistic logic and its measures of information and conflict. Fundamenta Informaticae **57**, 1–25 (2003)
77. Dunn, J.M.: Relevance logic and entailment. In: Gabbay, D., Gunthner, F. (eds.) Handbook of Philosophical Logic, vol. III, pp. 117–224. Riedel, Dordrecht (1986)
78. Elfes, A.: Using occupancy grids for mobile robot perception and navigation. Comput. Mag. **22**, 46–57 (1989)
79. Eytan, M.: Tableaux de Smullyan, ensebles de Hintikka et tour ya: un point de vue Algebriquem. Math. Sci. Hum. **48**, 21–27 (1975)
80. Fagin, R., Halpern, J., Moses, Y., Vardi, M.: Reasoning About Knowledge. MIT Press, Cambridge (Mass) (1995)
81. Fausett, L.: Fundamentals of Neural Network Architectures, Algorithms, and Applications. Prentice Hall, Upper Saddle River (1994)
82. Fitting, M.: Bilattices and the semantics of logic programming. J. Log. Program. **11**, 91–116 (1991)
83. Fitting, M.: A theory of truth that prefers falsehood. J. Philos. Log. **26**, 477–500 (1997)
84. Gabbay, D.: Labelled Deductive Systems, vol. 1. Oxford University Press, Oxford (1996)
85. Gentzen, G.: Collected Papers of Gerhard Gentzen (Szabo, M.E. (ed.)). North-Holland, Amsterdam (1969)
86. Ginsberg, M.: Multivalued logics. In: Proceedings of AAAI'86, pp. 243–247. Morgan Kaufman, Los Altos (1986)
87. Ginsberg, M.: Multivalued logics: a uniform approach to reasoning in AI. Comput. Intell. **4**, 256–316 (1988)
88. Halpern, J., Moses, Y.: Towards a theory of knowledge and ignorance: preliminary report. In: Apt, K. (ed.) Logics and Models of Concurrent Systems, pp. 459–476. Springer, Berlin (1985)
89. Halpern, J., Moses, Y.: A theory of knowledge and ignorance for many agents. J. Log. Comput. **7**, 79–108 (1997)
90. Hintikka, S.: Knowledge and Belief. Cornell University Press, Ithaca (1962)
91. Hughes, G., Cresswell, M.: An Introduction to Modal Logic. Methuen, London (1968)
92. Hughes, G., Cresswell, M.: A New Introduction to Modal Logic. Routledge, New York (1996)
93. Jaffar, J., Lassez, J.-L., Maker, M.: Some issues and trends in the semantics of logic programming. In: Shapiro, E. (ed.) Proceedings of the 3rd International Conference on Logic Programming. Lecture Notes Computer Science, vol. 225, pp. 223–241. Springer, Berlin (1986)
94. Jaffar, J., Stucky, P.: Canonical logic programs. J. Log. Program. **3**, 143–155 (1986)
95. Jaśkowski, S.: Propositional calculus for contradictory deductive systems (in Polish). Studia Societatis Scientiarun Torunesis, Sectio A **1**, 55–77 (1948)

96. Jaśkowski, S.: On the discursive conjunction in the propositional calculus for inconsistent deductive systems (in Polish). Studia Societatis Scientiarun Torunesis, Sectio A **8**, 171–172 (1949)
97. Kifer, M., Lausen, G., Wu, J.: Logical foundations of object-oriented and frame-based language. J. ACM **42**, 741–843 (1995)
98. Kifer, M., Lozinskii, E.L.: RI: a logic for reasoning with inconsistency. In: Proceedings of LICS4, pp. 253–262 (1989)
99. Kifer, M., Lozinskii, E.L.: A logic for reasoning with inconsistency. J. Autom. Reason. **9**, 179–215 (1992)
100. Kifer, M., Subrahmanian, V.S.: On the expressive power of annotated logic programs. In: Proceedings of the 1989 North American Conference on Logic Programming, pp. 1069–1089 (1989)
101. Kifer, M., Subrahmanian, V.S.: Theory of generalized annotated logic programming. J. Log. Program. **12**, 335–367 (1992)
102. Kifer, M., Wu, J.: A logic for object-oriented logic programming. In: Proceedings of 8th ACM SIGACT/SSIGMOD/SIGART Symposium on Principles of Database Systems, pp. 379–393 (1989)
103. Kleene, S.: Introduction to Metamathematics. North-Holland, Amsterdam (1952)
104. Kotas, J.: The axiomatization of S. Jaskowski's discursive logic. Studia Logica **33**, 195–200 (1974)
105. Kowalski, R.: Predicate logic as a programming language. In: Proceedings of IFIP'74, pp. 569–574 (1974)
106. Kowalski, R.: Logic for Problem Solving. North-Holland, Amsterdam (1974)
107. Kripke, S.: Outline of a theory of truth. J. Philos. **72**, 690–716 (1975)
108. Lee, R.: Fuzzy logic and the resolution principle. J. ACM **19**, 109–119 (1972)
109. Lewin, R.A., Mikenberg, I.F., Schwarze, M.G.: On the algebraizability of annotated logics. Studia Logica **59**, 359–386 (1997)
110. Lewin, R.A., Mikenberg, I.F., Schwarze, M.G.: Algebras and matrices for annotated logics. Studia Logica **65**, 137–153 (2000)
111. Lloyd, J.: Foundations of Logic Programming, 2nd edn. Springer, Berlin (1987)
112. Liu, X., Tsai, J., Weigert, T.: Λ-resolution and the interpretation of Λ implication in fuzzy operator logic. Inf. Sci. **56**, 259–278 (1991)
113. Lu, J., Henschen, L., Subrahmanian, V., da Costa, N.: Reasoning in paraconsistent logic. In: Boyer, R. (ed.) Automated Reasoning, pp. 181–210. Kluwer, Dordrecht (1991)
114. Lu, J.: Logic programming with signs and annotations. J. Log. Comput. **6**, 755–778 (1996)
115. Lu, J., Murray, N.V., Rosenthal, E.: Signed formulas and annotated logics. In: Proceedings of the 23rd International Symposium on Multiple-Valued Logic, pp. 48–53 (1993)
116. Łukasiewicz, J.: On 3-valued logic, 1920. In: McCall, S. (ed.) Polish Logic, pp. 16–18. Oxford University Press, Oxford (1967)
117. Mendelson, E.: Introduction to Mathematical Logic, 3rd edn. Wadsworth and Brooks, Monterey (1987)
118. McDermott, D.: Nonmonotonic logic II. J. ACM **29**, 33–57 (1982)
119. McDermott, D., Doyle, J.: Non-monotonic logic I. Artif. Intell. **13**, 41–72 (1980)
120. Minsky, M.: A framework for representing knowledge. In: Haugeland, J. (ed.) Mind-Design, pp. 95–128. MIT Press, Cambridge (Mass) (1975)
121. Moore, R.: Semantical considerations on nonmonotonic logic. Artif. Intell. **25**, 75–94 (1985)
122. Mortensen, C.: Every quotient algebra for C_1 is trivial. Notre Dame J. Form. Log. **21**, 694–700 (1980)
123. Mortensen, C.: Inconsistent Mathematics. Kluwer, Dordrecht (1995)
124. Murai, T., Akama, S., Kudo, Y.: Rough and fuzzy sets from a point of view of propositional annotated modal logic, Part 1. In: Proceedings of IFSA'2003, Istanbul, Turkey, pp. 241–244 (2003)
125. Nakamatsu, K.: On the relation between vector annotated logic programs and defeasible theories. Log. Log. Philos. **8**, 181–205 (2000)

126. Nakamatsu, K., Akama, S., Abe, J.M.: An intelligent coordinated traffic signal control based on EVALPSN. In: Knowledge-Based Intelligent Information and Engineering Systems: Proceedings of KES 2007—WIRN 2007, Part II. Lecture Notes on Artificial Intelligence, vol. 4693, pp. 869–876. Springer, Berlin (2007)
127. Nakamatsu, K., Akama, S., Abe, J.M.: Real-time intelligent process order control based on a paraconsistent annotated logic program EVALPSN. In: Proceedings of SSD'08 (2008)
128. Nakamatsu, K., Akama, S., Abe, J.M.: Application of Bf-EVALPSN to real-time process order control. In: Proceedings of BICS2008 (2008)
129. Nakamatsu, K., Suzuki, A.: Annotated semantics for default reasoning. In: Dai, R. (ed.) Proceedings of Pacific Rim International Conference on Artificial Intelligence (PRICAI'94), Beijing, China, pp. 180–186 (1994)
130. Nakamatsu, K., Suzuki, A.: A nonmonotonic ATMS based on annotated logic programs. In: Boecke, W. et al. (eds.) Agents and Multi-Agents Systems. LNAI, vol. 1441, pp. 79–93. Springer, Berlin (1998)
131. Negoita, C., Ralescu, D.: Applications of Fuzzy Sets to Systems Analysis. Wiley, New York (1975)
132. Nelson, D.: Constructible falsity. J. Symb. Log. **14**, 16–26 (1949)
133. Nelson, D.: Negation and separation of concepts in constructive systems. In: Heyting, A. (ed.) Constructivity in Mathematics, pp. 208–225. North-Holland, Amsterdam (1959)
134. Ng, R., Subrahmanian, V.S.: Probabilistic logic programming. Inf. Comput. **101**, 150–201 (1992)
135. Prawitz, D.: Natural Deduction: A Proof-Theoretical Study. Almqvist and Wiksell, Stockholm (1965)
136. Priest, G., Routley, R., Norman, J. (eds.): Paraconsistent Logic: Essays on the Inconsistent. Philosopia Verlag, München (1989)
137. Priest, G.: Logic of paradox. J. Philos. Log. **8**, 219–241 (1979)
138. Priest, G.: Paraconsistent logic. In: Gabbay, D., Guenthner, F. (eds.) Handbook of Philosophical Logic, 2nd edn., pp. 287–393. Kluwer, Dordrecht (2002)
139. Priest, G.: In Contradiction: A Study of the Transconsistent, 2nd edn. Oxford University Press, Oxford (2006)
140. Reiter, R.: A logic for default reasoning. Artif. Intell. **13**, 81–132 (1980)
141. Restall, G.: An Introduction to Substructural Logics. Routledge, London (2000)
142. Rico, G.O.: The annotated logics OP_{BL}. In: Carnielli, W., Coniglio, M., D'Ottaviano, I. (eds.) Paraconsistency: The Logical Way to the Inconsistent, pp. 411–433. Marcel Dekker, New York (2002)
143. Robinson, J.A.: A machine-oriented logic based on the resolution principle. J. ACM **12**, 23–41 (1965)
144. Routley, R., Plumwood, V., Meyer, R.K., Brady, R.: Relevant Logics and Their Rivals, vol. 1. Ridgeview, Atascadero (1982)
145. Shoenfield, J.: Mathematical Logic. Addison Wesley, Reading (1967)
146. Shafer, G.: A Mathematical Theory of Evidence. Princeton University Press, Princeton (1967)
147. Slaney, J.K.: A general logic. Australas. J. Philos. **68**, 74–88 (1990)
148. Smullyan, R.: First-Order Logic. Springer, Berlin (1969)
149. Subrahmanian, V.: On the semantics of quantitative logic programs. In: Proceedings of the 4th IEEE Symposium on Logic Programming, pp. 173–182 (1987)
150. Sylvan, R., Abe, J.M.: On general annotated logics, with an introduction to full accounting logics. Bull. Symb. Log. **2**, 118–119 (1996)
151. Thirunarayan, K., Kifer, M.: A theory of nonmonotonic inheritance based on annotated logic. Artif. Intell. **60**, 23–50 (1993)
152. Torres, C.R., Reis, R.: The new hardware structure of the Emmy II robot. In: Abe, J.M. (ed.) Paraconsistent Intelligent Based-Systems. Springer (to appear)
153. van Emden, M.: Qualitative deduction and fixpoint theory. J. Log. Program. **4**, 37–53 (1986)
154. van Emden, M., Kowalski, R.: The semantics of predicate logic as a programming language. J. ACM **23**, 733–742 (1976)

155. van Fraassen, B.C.: Facts and tautological entailment. J. Philos. **66**, 477–487 (1969)
156. Vasil'ev, N.A.: Imaginary Logic. Nauka, Moscow (1989) (in Russian)
157. Wansing, H.: The Logic of Information Structures. Springer, Berlin (1993)
158. Weigert, T., Tsai, J., Liu, X.: Fuzzy operator logic and fuzzy resolution. J. Autom. Reason. **10**, 59–78 (1993)
159. Zadeh, L.: Fuzzy sets. Inf. Control **8**, 338–353 (1965)
160. Zadeh, L.: Fuzzy sets as a basis for a theory of possibility. Fuzzy Sets Syst. **1**, 3–28 (1976)

Index

A
Abnormality, 107
Abstract logic, 85
Adaptative logic, 107
Adaptive, 107
Adaptive strategy, 107
Algebraic semantics, 31
Algebraizable, 85
α-model, 176
Annotated atom, 5, 25, 124
Annotated clause, 125
Annotated function, 124
Annotated literal, 112
Annotated logic program with strong negation, 138
Annotated logics, vii, 2
Annotated modal logic, 53
Annotated model theory, 40
Annotated set theory, 37
Annotation, 112, 124
Annotation variable, 124
Antinomy, 104
Approximate reasoning, 61
Artificial Neural Network, 154
Artificial neuron, 154
Autoepistemic logic, 139
Automated theorem-proving, vii
Axiom, 13

B
Basic PANC, 157
Bilattice, 109
Bilattices, 95
Body, 112, 125
Boolean-valued model, 176

C
Canonical, 119
Canonical Kripke Model, 57
Canonical merge, 150
Chain, 46
Classic, 32
Classical logic, vii, 1
Clausal form, 144
Closure, 41
Complementary property, 8
Complex annotation term, 124
Complex elements, 32
Complex formula, 6, 26
Confidence, 71
Confident function, 73
Conflation, 110
Consistent, 2, 17
Constrained clause, 131
Constrained query, 130
Constructive logic with strong negation, 105
Continuous, 117
C-system, 3, 99
Curry algebra, 31, 33
Curry pre-ordered system, 32
Curry systems, 31

D
Deduction, 133, 151
Deduction theorem, 15, 70
Default logic, 139
Defining equation, 85
Dempster-Shafer theory, 177
Denotation, 26
Deontic logic, 1
Diagram, 42
Dialetheism, 107

J.M. Abe et al., *Introduction to Annotated Logics*,
Intelligent Systems Reference Library 88, DOI 10.1007/978-3-319-17912-4

Discursive logic, 3, 98
Domain, 26
Downarrow iteration, 116
Dynamic dialectical logic, 107

E
EEG, 160
Elementary chain, 46
Elementary equivalent, 44
Elementary extension, 45
Elementary structure, 45
Elimination, 151
Elimination rule, 48
Emmy, 162, 165
Emmy III, 171
Epistemic logic, 1
Epistemic negation, 127
Equivalence, 6, 26
$E\tau$, 154
Existential quantifier, 25
Extension, 44
External dynamics, 107

F
Factor, 149
Factoring, 149
Falsity pattern, 158
F-formula, 141
Filter, 34
Finite annotation property, 22
First-degree entailment, 107
F-logic, 139
Flou set, 40
Four-valued logic, 102
Frame, 55, 139
F-structure, 141
Fuzzy annotated logics, 61
Fuzzy formula, 62
Fuzzy logic, 61
Fuzzy operator logic, 61
Fuzzy set, 39

G
General annotated logics, 98
General Herbrand interpretation, 126
General logics, 95
Generalized annotated logic programming, 124
Generalized annotated logics, 4, 23
Generalized annotated program, 125
Generalized Horn clause, 112
Graded modus ponens, 66
Graded paraconsistency, 67
Ground, 147

H
Head, 112, 125
Herbrand base, 126, 145
Herbrand model, 145
Herbrand structure, 145
Herbrand universe, 140, 145
Herbrand's theorem, 146
Hilbert system, 13
Hyper-literal, 6, 26
Hyper-literal elements, 32

I
Ideal, 35, 125
Id-term, 140
Imaginary logic, 3, 104
Implicative pre-lattice, 32
Inconsistency-adaptive logic, 107
Inconsistent, 2, 17, 42
Individual, 42
Inductive annotated logic, 71
Input restriction, 150
Instance, 147
Intensional logics, 1
Internal dynamics, 107
Interpretation, 6, 26
Introduction rule, 48
Intuitionistic logic, 1
IS-A acyclicity, 149
Is-a assertion, 140
IS-A reflexivity, 149
IS-A transitivity, 149
Isomorphism, 43

K
Knowledge representation, 138
Kripke model, 55
Kripke semantics, 55

L
Labelled Deductive Systems, 95
Language, 5
Lattice of truth-values, 5
Learning factor, 158
L-fuzzy set, 40
Literal, 112
Logic, vii
Logic of paradox, 105

Logic programming, vii, 111
Logically equivalent, 41
Logically valid, 42
Logics of Formal Inconsistency, 108
Lower limit logic, 107

M
Many-valued logic, 1
Matrix, 40
Merge, 150
Merging, 150
Modal logic, 1, 53
Model, 8, 27, 143
Model-theoretic semantics, 6
Modus ponens, 14
Molecular formula, 140
Monadic Curry algebra, 37
Most general unifier, 147

N
Natural capacity of learning, 158
Natural deduction, 48
Nave set theory, 177
Necessitation, 54
Necessity measure, 65
Necessity operator, 53
Necessity-valued formula, 65
¬-inconsistent, 9
Neural computing, 154
Nh-resolvent, 121
Non-alethic, 10
Non-alethic logic, 3
Non-classical logics, vii, 1
Non-classical world, 56
Non-monotonic logic, 139
Non-monotonic reasoning, 107, 139
Non-trivial, 2, 9, 17, 42
Nonstrict, 140
Normal, 131
Normal structure, 30

O
Object, 139
Object constructor, 140
Object molecule, 140
Object-oriented, 139
Object-oriented programming, 138
Ontological negation, 127
Open, 40
Output relaxation, 150

P
Paracomplete, 2, 9, 43
Paracomplete logic, 3
Paraconsistent, 9, 43
Paraconsistent Artificial Neural Cell of Learning, 158
Paraconsistent Artificial Neural Networks, 154
Paraconsistent logic, vii, 2, 98
Paraconsistent logic programming, vii, 111
Paraconsistent logical controller, 162
Paralog, 138
ParaLog-e, 138
Paramodulation, 149
Parasonic, 163
Pierce's law, 14
Possibilistic logic, 64, 109
Possibility distribution, 65
Possibility measure, 65
Possibility operator, 54
Possibility theory, 65
Possible world, 55
Pre-bilattice, 109
Pre-Boolean algebra, 33
Pre-lattice, 32
Predicate molecule, 144
Prefix, 40
Prenex form, 40
Principle of excluded middle, 3
Principle of non-contradiction, 2
Probabilistic logic programming, 177
Prolog, 111
p-tuple, 144

Q
Quantum computing, 177
Query, 121, 130

R
Range, 9
Reductant, 129
Refutation, 133, 151
Regular, 45
Relevance logic, 3, 100
Resolution, 111, 148
Restricted Herbrand interpretation, 126
Routley-Meyer semantics, 100
Rule, 48
Rules of inference, 13

S
Semantic consequence, 8, 27, 42
Semantic network, 139
Semantics, 6
Sequent calculus, 48
Setup, 103
Signed formula logic programming, 138
S-instance, 42
SLDnh-deduction, 122
SLDnh-refutation, 122
SLDnh-resolution, 121
Solvable, 131
Strong negation, 6, 26, 105
Strongly transitive, 38
Structural, 74
Structure, 41
Subclass inclusion, 149
Submodel, 44
Submolecule, 147
Substitution, 147
Substructural logic, 176
Substructure, 44
Syllogistic, 3
Syntactic consequence relation, 14

T
Temporal logic, 1
Term, 25
Theorem, 14
Theory of confidence, 73
Three-valued logic, 3
Threshold of certainty, 62
Transitive set, 38
Trivial, 2, 17
Truth pattern, 158
Truth relation, 27
Type inheritance, 150

U
Ultrafilter, 35
Unifiable, 112
Unification, 147
Unifier, 147, 148
Union, 46, 47
Universal-existential, 41
Universal quantifier, 25
Universe, 38, 42
Unrestricted deduction, 133
Unrestricted resolvent, 131
Upper limit logic, 108
Upward iteration, 116

V
Vagueness, 71
Valid, 8, 27, 42
Valuation, 6
Variable assignment, 26, 142
Vector annotated logic program with strong negation, 138

W
Warning rule, 73
Weakly continuous, 117
Well-behaved, 117

Printed in the United States
By Bookmasters